浙江省建筑设计研究院作品选

SELECTED WORKS
2010—2013

图书在版编目（CIP）数据

浙江省建筑设计研究院作品选：2010~2013 / 浙江省建筑设计研究院作品选编委会编著. -- 天津：天津大学出版社，2014.6
 ISBN 978-7-5618-5098-5

 Ⅰ. ①浙… Ⅱ. ①浙… Ⅲ. ①建筑设计－作品集－中国－现代 Ⅳ. ①TU206

中国版本图书馆CIP数据核字(2014)第131380号

责任编辑　王　尧
装帧设计　蔡金福
文字翻译　王　坤
摄　　影　吴　丹　蔡金福
策　　划　杭州衡安文化艺术策划有限公司

出版发行　天津大学出版社
出 版 人　杨欢
地　　址　天津市卫津路92号天津大学内（邮编：300072）
电　　话　发行部 022-27403647
网　　址　publish.tju.edu.cn
印　　刷　深圳市新视线印务有限公司
经　　销　全国各地新华书店
开　　本　220 mm ×300 mm
印　　张　19.75
字　　数　415千
版　　次　2014年8月第1版
印　　次　2014年8月第1次
定　　价　298.00元

凡购本书，如有质量问题，请向我社发行部门联系调换

《浙江省建筑设计研究院作品选2010—2013》编委会

主　　任：施祖元
副 主 任：曹跃进　许世文
学术顾问：益德清　唐葆亨　景政治
　　　　　方子晋
执行主编：蔡金福
编　　委：李志飙　杨学林　蒋　纹
　　　　　黄志斌　蔡　彤　吴　丹

简　介

浙江省建筑设计研究院创立于1952年，是浙江省最早成立的设计院，系浙江省住房和城乡建设厅直属事业单位，实行企业化管理，是一家具有良好社会声誉的综合性建筑设计与咨询机构。

本院以"设计创新，质量创优，诚信求实，团结敬业，发展争先"为宗旨。服务范围包括：建筑工程设计与咨询、城乡规划编制、建筑智能化系统工程设计、室内装饰设计、岩土工程设计、市政公用工程设计与咨询、风景园林设计、工程项目可行性研究、项目评估、工程造价咨询、工程项目管理和工程总承包等领域。

本院现共有职工503人，其中各类专业技术人员466人，中国工程设计大师2人，浙江省工程勘察设计大师3人，浙江省有突出贡献中青年专家2人，教授级高级工程师50人，高级建筑师和高级工程师163人，国家一级注册建筑师83人，一级注册结构工程师79人，注册规划师13人，其他各类注册资格人员100人。

本院专业齐全，设有9个综合建筑设计所，并设有规划与城市建筑研究所、建筑智能化设计研究所、绿色建筑工程设计研究中心、结构与岩土工程研究室、建筑经济室、装饰设计分院、市政设计分院、景观设计分院、建筑幕墙分院9个专业设计所（中心）、室、分院，9个职能部门以及浙江省建筑设计研究院技术发展中心、浙江建院工程咨询有限公司、浙江华之建筑设计有限公司、浙江华智建筑技术推广中心等4个子公司；另设有上海分院和温州、台州办事处。

本院坚持创新发展的设计理念，在办公、宾馆、体育、交通、文化、教育和医疗建筑设计方面具有明显优势，在复杂超限高层结构、大跨度空间结构和岩土工程设计等方面处于行业领先地位。本院下属的技术发展中心是浙江省具有超限高层建筑施工图设计文件审查资质的机构。

本院重视科技进步和技术创新，建院以来，累计荣获国家、省部级各类奖项400余项，科研成果获省部级科学技术奖20余项；主编、参编了行业标准《办公建筑设计规范》、浙江省标准《建筑地基基础设计规范》等30多项国家、行业和地方标准。

本院于1997年通过ISO9001质量管理体系认证，荣获"全国优秀勘察设计院"、首批"全国守合同重信用企业"、"'十五'全国建设科技先进集体"和"浙江省文明单位"等多项荣誉称号。

2014年6月1日

Brief Introduction

Founded in 1952, Zhejiang Province Institute of Architectural Design and Research (ZIAD) is the earliest design institute in Zhejiang Province. ZIAD is an institute unit directly affiliated to Bureau of Housing and Urban-Rural Development of Zhejiang Province. It adopts enterprise management and is a comprehensive architectural design and consulting organization enjoying high social reputation.

ZIAD has been insisting on the tenet of "innovative design, optimal quality, faithfulness, factualism, unity and competitiveness" and engages themselves in architectural engineering design and consultation, compilation of urban and rural planning, engineering design of architectural intelligent system, interior decorative design, geotechnical engineering design, municipal public facility design and consultation, landscape design, engineering feasibility research, project appraisal, construction cost consultation, project management and general contract, etc.

ZIAD has 503 employees in total, including 466 technicians, 2 national engineering designers, 3 provincial engineering survey designers, 2 provincial outstanding contribution young experts, 50 professor-level senior engineers, 163 senior architects and senior engineers, 83 national class I registered architects, 79 class I registered structural engineers, 13 registered planners and 100 other registered professionals.

ZIAD owns 9 comprehensive architectural design institutes and 9 professional design institutes (centers), departments or branches, including planning and urban architectural research institute, intelligent architecture design and research institute, green building engineering design and research center, structural and geotechnical engineering department, architectural economic department, decorative design branch, municipal design branch, landscape design branch and building curtain wall branch, and 9 functional departments, as well as four subsidiaries, including ZIAD Technical Development Center, ZIAD Engineering Consulting Co., Ltd., Zhejiang Huazhi Architecture Design Co., Ltd., Zhejiang Huazhi Architectural Technique Promotion Center, etc. In addition, ZIAD also sets a branch in Shanghai and offices in Wenzhou and Taizhou.

ZIAD insists on the design concept of innovative development, having great advantages on the aspects of the design of office building, hotel building, sports building, traffic building, cultural building and medical building, etc., and leading the industry on the aspects of design of complex over-limit high-rise structure, wide-span space structure and geotechnical engineering design, etc. ZIAD Technical Development Center is the organization in Zhejiang Province having the qualification for examination of construction drawing design document of over-limit high-rise building.

ZIAD encourages technological improvement and technical innovation, having won more than 400 national, provincial and ministry-level prizes and more than 20 provincial and ministry-level scientific and technical prizes, drafted and participated in drafting more than 30 national, professional and provincial standards, such as professional standard of Design Code for Office Building and provincial standard of Code for Design of Building Foundation, etc.

ZIAD has obtained ISO9001 Quality Management System Certificate in 1997 and has won many honorary titles, including National Excellent Survey and Design Institute, the First National Contract and Credit Respected Enterprises, the 10th Five-Year Plan National Scientifically and Technically Advanced Construction Group, and Civilized Unit of Zhejiang Province, etc.

2014-6-1

目　录

综合建筑
16　杭州创新创业新天地文创园
18　鸿发国际广场方案
20　慈溪市观海卫镇新城核心区公建群
22　民和·惠风和畅文化产业园
24　诺尔康上海奉贤经济开发区生产基地
26　杭州金融国际会展中心
28　英冠·世纪财富中心
30　宁波东部新城B1—5地块
31　天台（D2—2—3）地块项目
32　德清开元森泊度假村中央设施
34　萧山浪琴湾（公建部分）
35　杭州恒兴大厦
36　丽晶国际中心
38　慈溪财富中心
39　迪安诊断产业基地建设项目
40　慈溪市客运中心站新建工程
42　乌什县社会综合福利中心
44　浙江国际影视中心综合大楼
46　香山四季中心
48　浦江县广播电视中心迁建项目一期工程
49　杭州临平永安·金鑫大厦
50　慈溪文化商务区03地块
52　乐清总部经济园
54　天津德力置业
55　常州亚西亚影城
56　杭州萧山国际机场T3航站楼
58　华能长兴电厂去工业化设计
59　杭州热电厂地块（局部）概念性方案设计

办公建筑
62　钱江世纪城J—01—03地块项目
64　衢州市社会保障服务中心
66　新疆阿拉尔市行政中心办公楼
68　新疆阿拉尔市法院、检察院
70　交通银行股份有限公司宁波分行大厦
71　山东东营市广饶县颐和国际大厦
72　超威电源有限公司新能源大厦
74　衢州国土资源局办公大楼
75　长兴电力调度及生产运检用房
76　湖州市吴兴区民防建设大楼
78　厦门福兴国际中心
79　黄岩区政府综合服务中心
80　杭州丰盛·恒泰广场
82　海正药业（杭州）有限公司总部大楼

Contents

Multi-Functional Building
- 16 Hangzhou New World Cultural Innovation Park
- 18 Scheme of Hongfa International Square
- 20 Public Buildings in the Core Area of Guanhaiwei Town, Cixi City
- 22 Minhe • Huifenghechang Cultural Industry Park
- 24 Nurotron Shanghai Fengxian Economic Development District Production Base
- 26 Hangzhou Financial International Convention and Exhibition Centre
- 28 Yingguan • Century Fortune Plaza
- 30 New Town B1-5 Area at the East of Ningbo
- 31 Tiantai Lot (D2-2-3) Project
- 32 Central Facility of Deqing Kaiyuan Senbo Holiday Resort
- 34 Xiaoshan Langqin Bay (Public Building Part)
- 35 Hangzhou Hengxing Building
- 36 Regent International Centre
- 38 Cixi Fortune Centre
- 39 Di'an Diagnostics Industrial Base
- 40 Cixi Passenger Transportation Center Station
- 42 Wushi Social General Welfare Center
- 44 Zhejiang International Film and TV Center General Building
- 46 Xiangshan Season Center
- 48 Pujiang County Broadcasting and Television Center Relocation Project Phase I
- 49 Hangzhou Linping Yong'an • Jinxin Mansion
- 50 Cixi Cultural Business District Plot 03
- 52 Yueqing Headquarters Economic Park
- 54 Tianjin Deli Real Estate
- 55 Asia Film Center in Changzhou
- 56 Xiaoshan International Airport T3 Terminal Building, Hangzhou
- 58 Huaneng Changxing Power Plant De-industrialization Design
- 59 Hangzhou Thermal Power Plant (Partial) Concept Plan

Office Building
- 62 Qianjiang Century CBD Plot J-01-03
- 64 Quzhou Social Security Service Center
- 66 Xinjiang Alaer Administrative Center Office Building
- 68 Xinjiang Alaer Municipal Court and Procuratorate
- 70 Bank of Communication Ningbo Branch Mansion
- 71 Guangrao County Yihe International Mansion, Dongying City, Shandong
- 72 New Energy Tower of Chilwee Power Co., Ltd.
- 74 Quzhou State Land Resource Bureau Office Building
- 75 Changxing Power Bureau Building for Power Dispatching and Production Maintenance
- 76 Wuxing District Civil Defense Construction Building, Huzhou City
- 78 Xiamen Fuxing International Center
- 79 Huangyan District Governmental General Service Center
- 80 Hangzhou Fengsheng • Hengtai Plaza
- 82 Headquarters of Hisun Pharmaceutical (Hangzhou) Co., Ltd.

	83	衢州建设局办公大楼
	84	湖州电力生产运维基地
	86	世华大厦
	87	慈溪龙山镇灵峰路综合写字楼
	88	诸暨城东中心区A2地块综合办公大楼
	90	正凯中心
	92	三宏国际大厦
	94	杭州和茂大厦
	95	麦道大厦
	96	台州中央商务区3—02，04地块设计
	98	阿里巴巴B2B二号园区
	100	杭州汇鑫大厦
	102	国金中心
	103	杭州华联钱塘会馆
	104	温岭九龙商务中心（办证中心）
	105	青川县行政中心
	106	宁波银行总部大厦
	108	钱江新城金融中心
	109	杭州日信国际中心
	110	潮峰钢构集团有限公司幕墙生产车间改造工程
	111	江西德兴市德兴大厦
	112	余姚商会大厦
	113	杭州协和大厦
	114	宁波银行象山大厦
	115	杭州经济技术开发区水质检测、调度、控制中心综合楼
酒店建筑	118	天台蓝海酒店
	119	台州方远大饭店
	120	海宁长安大酒店
	121	海口七星级产权式国际大酒店
	122	浙江国际影视中心综合服务大楼
	123	杭州西湖国宾馆后勤房
	124	湖北省农村信用社联合社培训中心
	126	金都海洋公园配套服务中心
	127	义乌日信国际大酒店
	128	中大·西郊半岛富春希尔顿酒店
	130	东方君悦
	131	横店国贸大厦会议中心
	132	天台山温泉度假山庄改扩建项目
	134	宿迁威尼斯假酒店
	136	绍兴柯桥嘉悦广场
	137	常州凯纳商务广场
	138	千岛湖润和度假酒店
商业建筑	142	遵义国际商贸城
	144	诸暨国际商贸城
	145	北海东盟国际商贸城
	146	重庆朝天门国际商贸城
	148	大连辽渔国际水产品市场
	149	五峰金东山家居建材城
	150	江苏涟水2011—15号宗地项目
	152	泰州泰茂城
	153	英冠·天地城
	154	湖北鄂州航宇国际商贸城
	155	杭州转塘狮子村商业项目
	156	杭州滨江76号A地块——世贸商务楼
	157	钱江世纪城富丽大厦

	83	Office Building of Quzhou Construction Bureau
	84	Huzhou Power Generation and Maintenance Base
	86	Shihua Tower
	87	General Office Building on Lingfeng Road of Longshan Town, Cixi
	88	General Office Building on Plot A2 of Eastern Central District, Zhuji
	90	Zhink Center
	92	Sanhong International Building
	94	Hangzhou Hemao Building
	95	Maidao Mansion
	96	Taizhou CBD Plots 3-02 and 04
	98	Alibaba B2B No.2 Park
	100	Hangzhou Huixin Mansion
	102	International Financial Center
	103	Hangzhou UDC Qiantang Chamber
	104	Wenling Jiulong Business Center (Certificate Service Center)
	105	Qingchuan Administration Center
	106	Bank of Ningbo Headquarters
	108	Qianjiang New Town Financial Center
	109	Hangzhou Rixin International Center
	110	Curtain Wall Production Plant Reconstruction of Triumpher Steel Structure Group Co., Ltd.
	111	Jiangxi Dexing Mansion
	112	Yuyao Chamber of Commerce Building
	113	Hangzhou Xiehe Building
	114	Bank of Ningbo Xiangshan Building
	115	Water Quality Inspection, Dispatching and Control Center General Building in Hangzhou Economic and Technological Development Zone
Hotel Building	118	Tiantai Lanhai Hotel
	119	Taizhou Fangyuan Hotel
	120	Haining Chang'an Grand Hotel
	121	Haikou Seven-star Condo International Hotel
	122	Zhejiang International Movie & Television Center Comprehensive Service Building
	123	The Rear Service Room of Hangzhou West Lake State Guest House
	124	Training Center of Hubei Province Rural Credit Cooperatives Union
	126	Supporting Service Center of Jindu Haiyang Park
	127	Yiwu Rixin International Hotel
	128	Zhongda • Xijiao Peninsula Fuchun Hilton Hotel
	130	Grand Hyatt Hotel
	131	Conference Center of Hengdian International Trade Building
	132	Renovation and Extension Project of Tiantai Mountain Hot Spring Resort
	134	Suqian Venice Resort Hotel
	136	Shaoxing Keqiao Jiayue Square
	137	Changzhou Kaina Business Square
	138	Qiandao Lake Runhe Resort Hotel
Commercial Building	142	Zunyi International Trade City
	144	Zhuji International Trade City
	145	Beihai ASEAN International Trade City
	146	Chongqing Chaotianmen International Trade City
	148	Dalian Liaoyu International Aquatic Product Market
	149	Wufeng Jindongshan Home Furnishing and Building Materials City
	150	Jiangsu Lianshui 2011-15 Parcel Project
	152	Taizhou Taimao City
	153	Yingguan • Tiandi City
	154	Hubei Ezhou Hangyu International Trade City
	155	Commercial Project of Shizi Village, Zhuantang Town, Hangzhou City
	156	Lot A, No.76 Binjiang, Hangzhou—Shimao Business Building
	157	Fuli Building of Qianjiang Century CBD

	158	杭州江口股份经济合作社等3家经合社商业综合用房
	159	杭州方家畈股份经济合作社商业用房
	160	昆明螺蛳湾国际商贸城
	162	杭州厚仁商业街
	163	宁波江北万达广场

文教建筑	166	温州大学瓯江学院
	168	上海视觉艺术学院国际艺术大师中心
	169	余杭区教师进修学校迁建工程
	170	杭州师范大学仓前校区二期C区
	172	浙江省信息化测绘创新基地
	174	慈溪龙山中学迁建工程
	175	乐清市实验小学迁建工程
	176	慈溪中学
	178	阿克苏地区中等职业技术学校
	179	乐清市乐成镇第七小学滨海校区
	180	乐清市乐成镇第五中学
	182	杭州师范大学湘湖校区
	184	浙江海洋学院萧山科技学院
	186	乐清文化中心
	187	遂昌城市文化综合体
	188	浙江省地质资料中心
	189	嘉兴市图书馆、博物馆二期工程
	190	杭州市河道陈列馆
	192	宁波市北仑区宁职院图书馆
	194	浙江档案馆新馆
	195	新疆和田影剧院

医院建筑	198	温州医学院附属第二医院瑶溪分院
	200	天台县人民医院迁建工程
	201	浙江省中医院国家中医临床研究基地科研综合楼
	202	平阳县人民医院异地扩建工程
	204	台州市立医院新院区
	205	苍南县人民医院迁建工程
	206	杭州市儿童医院医疗综合楼
	208	开化县人民医院门急诊综合楼
	209	浙江省新华医院门诊及急诊住院楼改造工程
	210	浙江大学医学院附属义乌医院
	212	东阳市人民医院医疗综合大楼
	214	北仑人民医院

住宅建筑	218	中铁·青秀城
	220	湖北荆门双仙社区
	222	临沂"联合·世纪新筑"B地块
	223	中大·杭州西郊半岛妙得阁
	224	黄岩东路安置房
	226	玉环县峦岩山小区
	227	山东临沂"旭徽·凤凰水城"
	228	杭州萧山潮闻天下
	230	库尔勒棉纺厂地块
	232	衢州保障性住房——锦西家园
	233	温岭利兹·水印华庭
	234	江干区笕桥黎明社区拆迁安置房项目
	236	荆门市凯凌·香格里拉小区三期
	238	贵阳市花溪区十和田项目

	158	Comprehensive Commercial Buildings of 3 Economic Cooperatives Including Hangzhou Jiangkou Share Economic Cooperative
	159	Commercial Buildings of Hangzhou Fangjiafan Share Economic Cooperative
	160	Kunming Luoshiwan International Trade City
	162	Hangzhou Houren Commercial Street
	163	Ningbo Jiangbei Wanda Plaza
Cultural and Educational Building	166	Oujiang College of Wenzhou University
	168	Shanghai Institute of Visual Art International Artists Center
	169	Relocation Project of Yuhang Teacher Training School
	170	Hangzhou Normal University Cangqian Campus Phase II Zone C
	172	Zhejiang Information-based Surveying and Mapping Innovation Base
	174	Relocation Project of Cixi Longshan Middle School
	175	Relocation Project of Yueqing Experimental Elementary School
	176	Cixi Middle School
	178	Akesu Secondary Vocational and Technical School
	179	Yuecheng No.7 Elementary School Binhai Campus, Yueqing City
	180	Yuecheng No.5 Middle School, Yueqing City
	182	Hangzhou Normal University Xianghu Campus
	184	Zhejiang Ocean University Xiaoshan Campus
	186	Yueqing Cultural Center
	187	Suichang Urban Cultural Complex
	188	Zhejiang Provincial Geoinformation Center
	189	Jiaxing Library and Museum Phase II
	190	Hangzhou Watercourse Exhibition Hall
	192	Beilun Ningbo Polytechnic Library, Ningbo City
	194	Zhejiang Provincial Archives New Building
	195	Xinjiang Hetian Theater
Medical Care Building	198	Yaoxi Branch of the Second Affiliated Hospital of Wenzhou Medical College
	200	Relocation Project of Tiantai County People's Hospital
	201	National Traditional Chinese Medicine Clinical Research Base Building in Zhejiang Provincial Hospital of Traditional Chinese Medicine
	202	Expansion Project of Pingyang County People's Hospital
	204	New Zone of Taizhou Municipal Hospital
	205	Relocation Project of Cangnan County People's Hospital
	206	Medical Building of Hangzhou Children's Hospital
	208	Outpatient Building of Kaihua County People's Hospital
	209	Outpatient Building Renovation Project in Zhejiang Xinhua Hospital
	210	Affiliated Yiwu Hospital of Medicine School, Zhejiang University
	212	General Medical Building of Dongyang People's Hospital
	214	Beilun People's Hospital
Residential Building	218	Zhongtie • Qingxiu City
	220	Shuangxian Community in Jingmen, Hubei
	222	Linyi "Lianhe • Excellent Life" Plot B
	223	Zhongda • Hangzhou West Suburb Peninsula Miaode Pavilion
	224	Huangyan Donglu Resettlement Housing Project
	226	Luanyan Mountain Residential District, Yuhuan County
	227	"Xuhui • Lake of Phoenix" in Linyi, Shandong
	228	Chaowen Tianxia in Xiaoshan, Hangzhou
	230	Korla Cotton Mill Lot
	232	Quzhou Indemnificatory Housing—Jinxi Garden
	233	Wenling Lizi • Shuiyin Garden
	234	Resettlement Houses of Liming Community in Jianqiao, Jianggan District
	236	Jingmen Kailing • Shangri-La Phase III Project
	238	Guiyang Huaxi Shihetian Project

	240	黄岩羽村安置房工程
	242	江苏汇丰城市综合体方案
	243	山东临沂"爱伦坡"
	244	杭州寰宇天下
	246	杭长铁路等重点工程农居拆迁安置房
	248	三门银轮玫瑰湾小区
	249	芜湖县南湖品园小区
	250	杭州阳光郡
	252	萧山浪琴湾
	254	山东淄博"正承·PARK"
	256	恒隆国际花园及恒隆国际酒店
	257	萧山云涛名苑
	258	山东郯城中央华庭
	260	当涂豪邦君悦华庭住宅小区
	261	浙江天台红石梁广场
体育建筑	264	临海市体育文化中心方案
	266	浙江工商大学文体中心
	268	奉化体育馆
	269	杭州市全民健身中心方案
	270	岱山县文化体育中心
	272	湖州南太湖湿地奥体公园
规划与城市设计	276	杭州未来科技城"城市绿心"规划
	278	杭州良渚组团中央商务区城市设计
	280	杭州未来科技城西溪科技岛城市设计
	282	湄洲湾职业技术学院迁建工程修建性详细规划
	284	嘉兴子城片区城市有机更新概念规划与城市设计
	286	杭州瓶窑老镇区概念规划
	288	嘉兴湖滨片区城市有机更新概念规划与城市设计
	290	杭州未来科技城重点建设区域城市色彩规划
	292	栖霞滨湖新城控制性详细规划及重点地段城市设计
	294	诸暨市枫桥镇枫江文化体育公园景观设计
幕墙设计	298	绍兴金沙半岛酒店
	300	杭州新天地E地块
	301	诸暨市规划展示馆和科技馆
	302	环球万豪国际中心
	303	海亮大厦
	304	海威银泰喜来登酒店
	305	杭州中威电子股份有限公司安防监控设备生产基地
	306	温州市江滨商务区CBD片区17—03#地块
	307	杭州新天地商务中心P4地块
	308	中信银行杭州分行总部大楼
	309	杭州钱江新城勇进中学
	310	上海长风8号东地块项目
	311	梦工场影视道具生产基地
	312	青山湖越秀城市综合体A区酒店
	313	乐清南虹广场

	240	Huangyan Yucun Village Resettlement Housing Project
	242	Jiangsu HSBC Urban Complex
	243	"Ailunpo" in Linyi, Shandong
	244	Hangzhou U World
	246	Resettlement Housing for Farmers for Major Projects like Hangzhou-Changsha Passenger-dedicated Line
	248	Sanmen Yinlun Rose Bay Residential District
	249	Nanhu Pinyuan Residential District in Wuhu County
	250	Hangzhou Sunny Estate
	252	Xiaoshan Langqin Bay
	254	"Zhengcheng•PARK" in Zibo, Shandong
	256	Henglong International Garden and Henglong International Hotel
	257	Xiaoshan Yuntao Garden
	258	Central Garden in Tancheng, Shandong
	260	Dangtu Haobang Junyuehuating Residential District
	261	Zhejiang Tiantai Hongshiliang Square
Sports Building	264	Design Scheme for Linhai Sports Culture Center
	266	Cultural and Sports Center in Zhejiang Gongshang University
	268	Fenghua Gymnasium
	269	Design Scheme for Hangzhou Citizen Fitness Center
	270	Daishan County Cultural and Sports Center
	272	Huzhou South Taihu Wetland Olympic Park
Plan and City Design	276	"Urban Green Center" Planning in Hangzhou Future Sci-Tech City
	278	Urban Design of Central Business District in Liangzhu, Hangzhou
	280	Urban Design of Xixi Sci-Tech Island in Hangzhou Future Sci-Tech City
	282	Construction Planning Details about Relocation of Meizhouwan Vocational Technology College
	284	Organic Renovation Conceptual Planning and Urban Design of Jiaxing Zicheng Plot
	286	Conceptual Planning of Pingyao Old Township of Hangzhou
	288	Urban Organic Renovation Conceptual Planning and Urban Design of Waterfront Plot in Jiaxing City
	290	Urban Color Planning at Key Construction Area of Hangzhou Future Sci-Tech City Project
	292	Controlled Planning Details and Urban Design of Important Locations of Qixia Waterfront New City Project
	294	Landscape Design of Fengqiao Town Fengjiang Cultural and Sports Park, Zhuji
Curtain Wall Design	298	Shaoxing Jinsha Peninsula Hotel
	300	Hangzhou New World Plot E
	301	Zhuji Planning and Exhibition Hall and Science and Technology Hall
	302	Huanqiu Marriott International Center
	303	Hailiang Tower
	304	Haiwei-Intime Sheraton Hotel
	305	Security Monitoring Equipment Production Base of OB Telecom Electronics Co., Ltd.
	306	Wenzhou Jiangbin Business District CBD Plot 17-03
	307	Hangzhou New World Business Center Plot P4
	308	China CITIC Bank Hangzhou Branch Headquarters
	309	Hangzhou Qianjiang Yongjin Middle School
	310	Shanghai Changfeng No.8 East Plot
	311	Dreamworks Film and TV Property Production Base
	312	Hotel in Qingshanhu Yuexiu Urban Complex Plot A
	313	Yueqing Nanhong Plaza

综合建筑
Multi-Functional Building

16	杭州创新创业新天地文创园
18	鸿发国际广场方案
20	慈溪市观海卫镇新城核心区公建群
22	民和·惠风和畅文化产业园
24	诺尔康上海奉贤经济开发区生产基地
26	杭州金融国际会展中心
28	英冠·世纪财富中心
30	宁波东部新城B1-5地块
31	天台（D2—2—3）地块项目
32	德清开元森泊度假村中央设施
34	萧山浪琴湾（公建部分）
35	杭州恒兴大厦
36	丽晶国际中心
38	慈溪财富中心

16	Hangzhou New World Cultural Innovation Park
18	Scheme of Hongfa International Square
20	Public Buildings in the Core Area of Guanhaiwei Town, Cixi City
22	Minhe • Huifenghechang Cultural Industry Park
24	Nurotron Shanghai Fengxian Economic Development District Production Base
26	Hangzhou Financial International Convention and Exhibition Centre
28	Yingguan • Century Fortune Plaza
30	New Town B1-5 Area at the East of Ningbo
31	Tiantai Lot (D2-2-3) Project
32	Central Facility of Deqing Kaiyuan Senbo Holiday Resort
34	Xiaoshan Langqin Bay (Public Building Part)
35	Hangzhou Hengxing Building
36	Regent International Centre
38	Cixi Fortune Centre

39	迪安诊断产业基地建设项目
40	慈溪市客运中心站新建工程
42	乌什县社会综合福利中心
44	浙江国际影视中心综合大楼
46	香山四季中心
48	浦江县广播电视中心迁建项目一期工程
49	杭州临平永安·金鑫大厦
50	慈溪文化商务区03地块
52	乐清总部经济园
54	天津德力置业
55	常州亚西亚影城
56	杭州萧山国际机场T3航站楼
58	华能长兴电厂去工业化设计
59	杭州热电厂地块（局部）概念性方案设计

39	Di'an Diagnostics Industrial Base
40	Cixi Passenger Transportation Center Station
42	Wushi Social General Welfare Center
44	Zhejiang International Film and TV Center General Building
46	Xiangshan Season Center
48	Pujiang County Broadcasting and Television Center Relocation Project Phase I
49	Hangzhou Linping Yong'an • Jinxin Mansion
50	Cixi Cultural Business District Plot 03
52	Yueqing Headquarters Economic Park
54	Tianjin Deli Real Estate
55	Asia Film Center in Changzhou
56	Xiaoshan International Airport T3 Terminal Building, Hangzhou
58	Huaneng Changxing Power Plant De-industrialization Design
59	Hangzhou Thermal Power Plant (Partial) Concept Plan

杭州创新创业新天地文创园
Hangzhou New World Cultural Innovation Park

本项目位于杭州市下城区创新创业新天地的西南角，为新天地的核心商业地段。地块西面紧靠东新东路绿化带，南面紧靠长大屋路并连接未来的地铁3号线，北临东文路和城市公园，东临工业遗存核心区。

项目地上建筑面积240 439 m²，包括5层商业裙房、2幢17层办公楼、2幢商务办公楼（分别为16层、17层）。地下建筑面积167 022 m²，为4层。整体建筑以横向的流畅线条把整个商业综合体的形象统一起来。横向线条的起伏与变化增加了商业综合体的韵律感，让建筑的形象更为突出。

This project is located in the southwest corner of the Innovation and Entrepreneurship New World in Xiacheng District, Hangzhou, which is the core business area of the Innovation and Entrepreneurship New World. This area is next to Dongxin East Road green belt in the west, south to Changdawu Road, connecting with Metro No. 3 in the future. It is close to Dongwen Road and City Park in the north, and the core area of industrial remains in the east.

The construction area above the ground is 240,439 m², including a 5-floor commercial podium, two 17-floor office buildings, two business office buildings (which have 16 floors and 17 floors respectively). the underground construction area is 167,022 m², which has 4 floors. The overall buildings apply horizontal smooth lines, which connect the whole commercial complex. The ups and downs and the change of the horizontal lines increase the rhythm of the commercial integration and make the image of the building more prominent.

建设单位：杭州利坤投资发展有限公司
功能用途：商业综合体
设计/竣工年份：2013年/~
建设地点：浙江省杭州市下城区
总建筑面积：407 500 m²
总建筑高度/层数：75 m/17 F
结构形式：框剪结构
合作单位：铿晓设计咨询（上海）有限公司

Owner: Hangzhou Likun Investment and Development Co., Ltd.
Function: Commercial Complex
Design/Completion Year: 2013/~
Construction Site: Xiacheng District, Hangzhou City, Zhejiang
Total Floor Area: 407,500 m²
Total Height/Floor: 75 m/17 F
Structure: Frame-Shear Wall Structure
Cooperation Unit: Hassell Design Consulting(Shanghai) Ltd.

鸿发国际广场方案
Scheme of Hongfa International Square

项目位于钱江世纪城H-08-2地块,规划建设用地面积为12 703 m²。方案总体布局上特意将主体塔楼放于地块东北角,让出西南面的大型城市公共空间,以期与文化景观轴和居住综合功能区完美衔接,并且给予足够的空间来迎接地铁2号线钱江世纪城站的交通人流和车流,同时还在东北角也构筑一个大型广场。主体塔楼两翼分别与公园东路和广场路平行,完美地与周边环境相互协调。办公主楼位于地块东北角并向西、南两向延伸展开,大型商场位于主楼西面。地块内围绕建筑形成环形消防车道,人流则由地块四边的四条道路进入。

This project is located in Qianjiang Century H-08-2 area. The designed construction area is 12,703 m². The overall layout of the project deliberately puts the main tower building in the northeast corner of the area, which makes the big city public area in the southwest connect perfectly with the cultural landscape and residence area. It also provides enough space to welcome the stream of people and transportation from the station of Qianjiang Century at Metro No. 2. In addition, a large square is going to be built in the northeast corner. The two wings of the main tower building are parallel with Park Road and Square Road, which can coordinate with the surroundings perfectly. The office building is located in the northeast corner of the area and extends to the southwest. The large shopping center is located at the west side of the main building. Within the area, a circular fire lane is formed around the building and the stream of people can enter inside through the four roads at four sides.

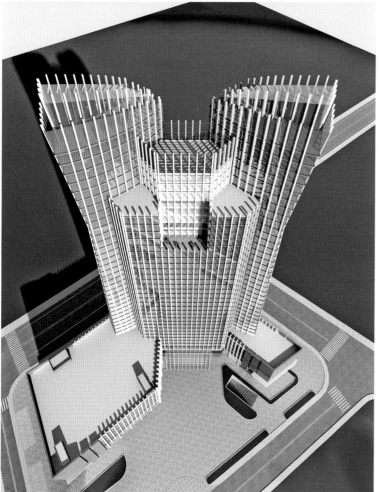

建 设 单 位：浙江鸿发置业有限公司
功 能 用 途：商业、办公、单身公寓
设计/竣工年份：2013年/~
建 设 地 点：浙江省杭州市
总建筑面积：76 200 m²
总建筑高度/层数：148.2 m/38 F
结 构 形 式：框筒结构

Owner: Zhejiang Hongfa Real Estate Co., Ltd.
Function: Commerce, Office, Single Apartment
Design/Completion Year: 2013/~
Construction Site: Hangzhou City, Zhejiang
Total Floor Area: 76,200 m²
Total Height/Floor: 148.2 m/38 F
Structure: Frame-Tube Structure

慈溪市观海卫镇新城核心区公建群
Public Buildings in the Core Area of Guanhaiwei Town, Cixi City

项目位于观海卫镇中部，是城市的主要门户节点和公共服务的核心区域，由商务办公中心、文化中心、体育中心、城市公园四个部分组成。规划中考虑到观海卫镇新城核心区的地理环境，设计以水为依托形成多元性活动绿廊，重建人与水的和谐关系。行政中心以院落式的布局围合而成，行政中心、文化中心、体育中心这三个建筑群体沿着中央水景布置，各方中心建筑相互呼应，使得建筑与环境共同生长，共同围合成一个崭新的新城核心区建筑群体。核心区的建筑将成为整个观海卫镇的城市中心，带动整个新城的开发和建设。

This project is located in the center of Guanhaiwei Town, which is the main gateway of the town and core area of public service. It consists of four parts, including business office center, cultural center, sports center, and city park. Considering the geographical environment of the core area of Guanhaiwei Town in the planning process, the multi-activity green gallery based on water is formed and the harmonious relationship between human and water is rebuilt in the design. The administrative center is in the form of courtyard. The administrative center, cultural center and sports center are arranged along the central water scenery. All the central buildings echo with each other, which could make the co-development of buildings and environment and form a new core area of buildings. The buildings in this core area will be the center of the entire Guanhaiwei Town and promote the development and construction of the new town.

建设单位：慈溪市观海卫镇人民政府
功能用途：综合
设计/竣工年份：2012年/～
建设地点：浙江省慈溪市观海卫镇
总建筑面积：139 080 m²
总建筑高度/层数：35.7 m/9 F
结构形式：框剪结构

Owner: Guanhaiwei Town People's Government of Cixi City
Function: Complex
Design/Completion Year: 2012/~
Construction Site: Guanhaiwei Town, Cixi City, Zhejiang
Total Floor Area: 139,080 m²
Total Height/Floor: 35.7 m/ 9 F
Structure: Frame-Shear Wall Structure

民和·惠风和畅文化产业园
Minhe · Huifenghechang Cultural Industry Park

项目位于宁波市国家高新区，总用地面积约 34 896 m²，拟建高层办公、商业建筑和动漫儿童体验馆及相关配套设施。最初的设计创意来自令人印象深刻的刺绣与皇帝龙袍上的"龙"的图案。"龙身"是由7栋办公楼组合而成，高低错落有致，好像龙在微妙地摆动。设计始终以最好的采光和通风来作为设计的基本要素。创意办公楼之间不仅留有足够的空间让中庭有足够的日照时间，还在楼与楼层之间专门设计了露天绿化平台。垂直景观是本方案不可分割的一部分，景观设计不仅仅停留在地面上，而且沿着地面景观和内部庭院，往垂直方向延伸，使露天平台形成一个个垂直的空中花园，让整座建筑充满绿色，使之具有独特的建筑形式。

This project is located in the National Hi-Tech Zone in Ningbo. The total area is 34,896 m². The proposed constructions include high-rise office, commercial buildings, animation museum for children, and the related facilities. The initial design ideas are from the impressive embroidery and the "dragon" pattern on the dragon robe. "The body of the dragon" consists of 7 office buildings, which are up and down, patchworks, just like a "dragon" in subtle swing. The design has been making sure the basic designing elements have best light and ventilation. There is not only enough space between the office buildings to allow enough daylight hours, but also an open-air green platform especially designed between the buildings. The vertical landscape of this project is an integral part and the design of the landscape does not only stop at the ground, but also extends to the vertical area along the ground landscape and internal courtyard, which makes the open platform a vertical garden in the air. It makes the whole building full of green and a unique style of architecture.

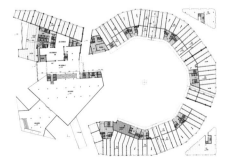
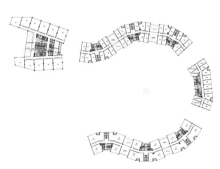

建设单位：宁波民和投资开发有限公司
功能用途：动漫体验、办公、商业
设计/竣工年份：2012年/~
建设地点：浙江省宁波市国家高新区
总建筑面积：120 000 m²
总建筑高度/层数：99.85 m/24 F
结构形式：框剪结构
合作单位：英国格林浩沃斯建筑设计事务所

Owner: Ningbo Minhe Investment Development Co., Ltd.
Function: Animation Experience, Office, Commerce
Design/Completion Year: 2012/~
Construction Site: National Hi-Tech Zone, Ningbo City, Zhejiang
Total Floor Area: 120,000 m²
Total Height/Floor: 99.85 m/ 24 F
Structure: Frame-Shear Wall Structure
Cooperation Unit: Glenn Howells Architects

诺尔康上海奉贤经济开发区生产基地
Nurotron Shanghai Fengxian Economic Development District Production Base

项目位于上海市奉贤区生物科技园区09-04地块，用地面积为73 196 m²。

方案在基地南侧中段沿汇丰北路设置主出入口，结合主出入口，在基地中心设计中心绿地广场，广场北侧为5层的厂房及管理用房，形成大气、开阔的基地南北主轴线。基地东侧为厂房一区和厂房二区，基地西侧为孵化生产厂房、孵化厂房和厂房三区。整个基地内各建筑与绿地景观有机结合，改善了整个基地内的生产环境。通过合理布置各功能建筑，生产所需的人流和物流将快捷方便并互不影响。

This project is located in the Biotechnology Park 09-04 area in Fengxian District, Shanghai, and covers an area of 73,196 m².

The main entrance and exit is designed to be at Huifeng North Road at the middle section of the south of the base. Combining with the main entrance and exit, the central green square is located in the center of the base. A 5-floor plant and management building are at the north of the square, which forms a main north–south axis of the base that is grand and wide. The first and second plant areas are in the east of the base. The west of the base is divided into hatchery production plant, hatchery plant and the third plant. The buildings and landscape in the base integrated with each other to improve the production environment of the whole base. Through the rational arrangement of the functional architecture, the people and goods for production can become efficient and convenient without interference.

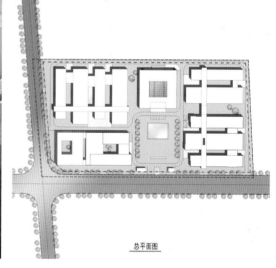

总平面图

建设单位：上海诺尔康神经电子科技有限公司
功能用途：生产厂房、办公、科研
设计/竣工年份：2011年/～
建设地点：上海奉贤经济开发区
总建筑面积：98 878 m²
总建筑高度/层数：23.9 m / 5 F
结构形式：框剪结构

Owner: Shanghai Nurotron Biotechnology Co., Ltd.
Function: Production Plant, Office, Scientific Research
Design/Completion Year: 2011/~
Construction Site: Economic Development Zone, Fengxian District, Shanghai City
Total Floor Area: 98,878 m²
Total Height/Floor: 23.9 m / 5 F
Structure: Frame-Shear Wall Structure

杭州金融国际会展中心
Hangzhou Financial International Convention and Exhibition Centre

项目位于杭州的西南部，东临钱塘江，西就绕城西线高速，北靠西湖风景名胜区山脉。设计将整个地块由西到东按功能分为A、B、C3个区，以会议展览为主，以旅游娱乐为辅，200 000余平方米的金融中心和200 000平方米的商业中心结合近80 000平方米的会展中心及包括4100个停车位等的配套设施，使之成为集金融、商业、会展于一体的综合服务中心。建筑在基地中均匀分布。建筑风格在新古典主义风格的基础上，推陈出新，充分借鉴古为今用、洋为中用的手法。整个建筑外形古朴庄重，线条鲜明，凹凸有致，呈现出一种华贵典雅的风格。外墙全部使用古朴典雅的干挂石材饰面，细节处理上运用了法式廊柱、雕花、线条，呈现出浪漫典雅的风格。

This project is located in the southwest of Hangzhou, with Qiantang River in the east, west high-way in the west, and the mountain areas of West Lake Scenic Area in the north. It is designed to divide the whole area into area A, B, C according to functions from west to east. The main function is for conference and exhibition and the subsidiary function is for tourism and entertainment. There are over 200,000 m² financial center and 200,000 m² commercial center with 80,000 m² exhibition center. In addition, there are also other facilities such as 4,100 parking spaces. This project can become the integrated service center for finance, business and exhibition. The buildings are arranged homogeneously at the base. The architectural style is based on neo-classical style and also innovative, which fully refers to ancient and western tactics. The shape of the whole building is simple and serious, with sharp lines and ups and downs, showing a luxurious and elegant style. The external wall applies simple and elegant stone finishes, using the French pillars, carving and lines for details, which show a romantic and elegant style.

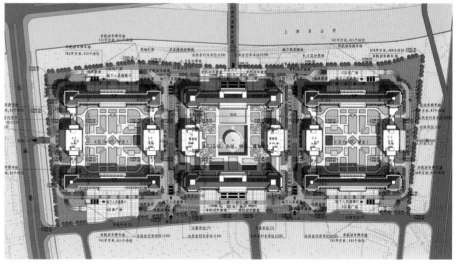

建设单位：杭州中融控股集团有限公司	Owner: Hangzhou Zhongrong Holding Group Co., Ltd.
功能用途：会议、办公、展览、商业、博览	Function: Conference, Office, Exhibition, Commerce, Expo
设计/竣工年份：2011年/~	Design/Completion Year: 2011/~
建设地点：浙江省杭州市西湖区	Construction Site: Xihu District, Hangzhou City, Zhejiang
总建筑面积：832 830 m²	Total Floor Area: 832,830 m²
总建筑高度/层数：54 m/12 F	Total Height/Floor: 54 m/12 F
结构形式：框剪结构	Structure: Frame-Shear Wall Structure
合作单位：法国DOMINIQUE HERTENBERGER ARCHITECTE DPLG	Cooperation Unit: France DOMINIQUE HERTENBERGER ARCHITECTE DPLG

英冠·世纪财富中心
Yingguan·Century Fortune Plaza

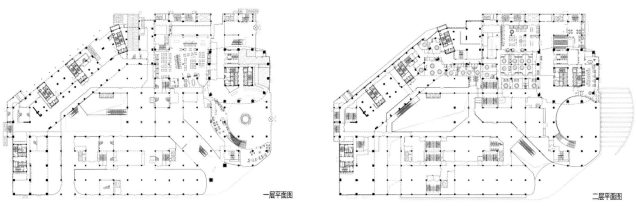

一层平面图　　　　二层平面图

设计将整个工程分为一个塔楼、两个板楼以及裙房，东侧为168 m高的五星级酒店，西北侧为甲级写字楼。一塔两板沿东、北、西三面围合布置，使得两幢楼的视觉干扰最小化。从主体至裙房，我们在规整柱网的前提下，采用有限度的弧线外墙将三者融为一体，形成流畅大气的空间形态和简洁的建筑形式，同时，又不失柔和灵动的韵味。与此同时，在三者的顶部又做了特殊的螺旋梭形造型，使建筑的丰富性、整体性达到和谐与统一，使得建筑形体在巨型体系中又含有活跃的因子，在震撼力中又透射出细腻、丰富的美感。

We divide the whole project into one tower building, two slab-type apartment building, and podiums. A 168 m five-star hotel will be in the east and a Grade-A office building will be in the northwest. One tower and two apartment will be laid out along the east, north and west, which minimizes the visual interference of the two buildings. From the main building to podiums, we apply limited arc exterior wall to integrate the three parts together on the premise of regular column grid to form a fluent and generous space, which forms a simple building with harmonious and flexible rhythm. At the same time, we also designed a special form of screw and spindle on the top of the three buildings, which unifies the richness, wholeness, and harmony of the buildings and makes the building an active factor in the giant building system. It shows delicate and rich beauty with a shocking power.

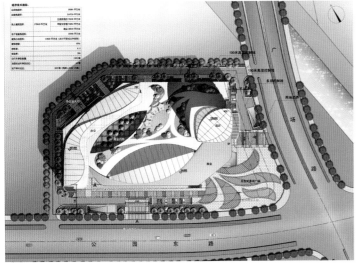

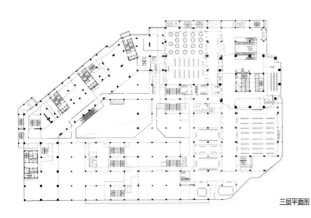

三层平面图

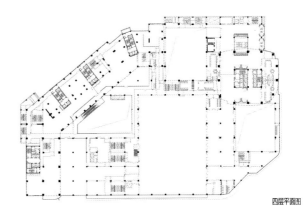

四层平面图

建设单位：	浙江中冠房地产开发有限公司	Owner: Zhejiang Zhongguan Real Estate Development Co., Ltd.
功能用途：	酒店、商业、办公	Function: Hotel, Commerce, Office
设计/竣工年份：	2011年/~	Design/Completion Year: 2011/~
建设地点：	浙江省杭州市萧山区	Construction Site: Xiaoshan District, Hangzhou City, Zhejiang
总建筑面积：	252650 m²	Total Floor Area: 252,650 m²
总建筑高度/层数：	150 m/42 F	Total Height/Floor: 150 m / 42 F
结构形式：	框架结构	Structure: Frame Structure

宁波东部新城B1—5地块
New Town B1-5 Area at the East of Ningbo

建 设 单 位：宁波东部新城开发投资有限公司
功 能 用 途：办公、商业
设 计/竣工年份：2011年/~
建 设 地 点：浙江省宁波市东部新城东区
总建筑面积：184 500 m²
总建筑高度/层数：116.6 m/26 F
结 构 形 式：框剪结构
合 作 单 位：朱锫建筑设计咨询（北京）有限公司

Owner: Ningbo Eastern New Town Development Investment Co., Ltd.
Function: Office, Commerce
Design/Completion Year: 2011/~
Construction Site: Eastern Area of Eastern New Town, Ningbo City, Zhejiang
Total Floor Area: 184,500 m²
Total Height/Floor: 116.6 m/ 26 F
Structure: Frame-Shear Wall Structure
Cooperation Unit: Studio Pei-Zhu

工程位于宁波市东部新城东区，城市重要礼仪性街道中山路以南。地块呈方形，项目规划目标是建设集总部办公、商务办公、休闲健身、商务餐饮、会务展览等功能于一体的商务办公综合体。项目用地面积为25 638 m²。建筑布局上，4座建筑相对独立又彼此联系，既灵活适应办公、商业的不同需求，又具备建筑间的良好互动关系。北侧两座写字楼通过裙房相系，南侧两座写字楼则在空中相连，使整组建筑形成了统一又独特的风格。

This project is located in the eastern area of Eastern New Town of Ningbo, and the south of the city's important ceremonial street— Zhongshan Road. The area is square. The aim of the project is to build an office and commercial complex of headquarters office, business office, leisure and fitness, business dining, conference and exhibition. The total area is 25,638 m². In terms of the layout, the four buildings are independent as well as connected with each other. It does not only meet the different requirements of office and business flexibly, but also has a sound interaction between the buildings. The two office buildings in the north are connected with podium, and the two buildings in the south are connected in the air, which makes a unique style of the whole building area.

天台（D2—2—3）地块项目
Tiantai Lot (D2-2-3) Project

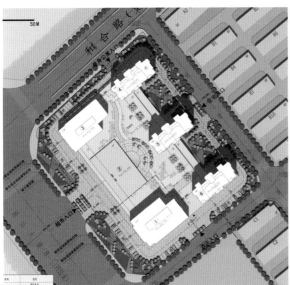

建设单位：天台县大户丁村村民委员会
功能用途：酒店、商业、住宅
设计/竣工年份：2011年/2014年
建设地点：浙江省天台县
总建筑面积：17 056.4 m²
总建筑高度/层数：59 m/19 F
结构形式：框剪结构

Owner: Dahuding Village Committee of Tiantai County
Function: Hotel, Commerce, Residence
Design/Completion Year: 2011/2014
Construction Site: Tiantai County, Zhejiang
Total Floor Area: 17,056.4 m²
Total Height/Floor: 59 m / 19 F
Structure: Frame-Shear Wall Structure

　　地块西侧设置超市入口广场，南侧设置酒店入口广场，北侧设置单身公寓入口，住宅入口则设置在地块东侧，并利用地块内环形车道与城市道路相连。交通流线明晰，各个功能出入口各自独立，互不干扰。酒店、单身公寓及超市单体建筑以玻璃幕墙、石材作为主要外墙材料，色彩上以浅色、中性色为主，结合深色玻璃，体现公共建筑现代简约的建筑风格。住宅单体以石材、玻璃和高级涂料作为主要外墙材料，色彩上以浅米色、偏暖色调为主，塑造时尚典雅、安宁温馨的高档住宅生活氛围。

The entrance of the supermarket is located in the west; the hotel entrance is in the south; the entrance of apartments for singles is in the north; the entrance of residential area is in the east. The project connects the circular road in the area with the city roads and the traffic lines are clear. All the entrances and exits are independent without interference. The main exterior materials of the hotel, apartments for singles, and supermarkets are glass wall and stone. The color is mainly light and neutral. Combining with dark glass, the project embodies the modern and simple construction style of public buildings. The main exterior materials for residential unit are stone, glass, and advanced coatings and the color is mainly light beige and warm-toned, creating a fashionable, elegant, peaceful and warm atmosphere of luxury residence.

德清开元森泊度假村中央设施
Central Facility of Deqing Kaiyuan Senbo Holiday Resort

本建筑为开元水上乐园度假区的核心建筑，是度假区的中央设施。项目集中了度假区的游乐、休闲、餐饮功能，并以水上乐园和探索冒险乐园为主体，借鉴了欧洲中央公园的游乐模式，仔细研究了中国及国外游客的游乐偏好，全力开发出一种新颖而富有魅力的游乐休闲体验模式，力求展现一种全新的度假游玩方式，全面提升德清开元度假区、杭州乃至中国的旅游文化品质与内涵，开创出一种全新的度假旅游业态。建筑设计的理念来自德清魅力山水和优秀的生态环境，将绿叶主题融入建筑造型，将环保主题贯穿建筑内外。

The building is the core building in Kaiyuan Water Park Holiday Resort, and is also the central facility of the holiday resort. The project is a concentration of recreation, relaxation and catering of the holiday resort, with Water Park and Adventure Park as main parts. It learns from the entertainment mode of European Central Park, does careful research to the entertainment preference of tourists from China and abroad and develops a new and glamorous entertainment and relaxation experience mode. It strives to show a brand new mode for holiday and entertainment, to promote the quality and connotation of tourism culture of Deqing Kaiyuan Holiday Resort, of Hangzhou and even China, and to create a brand new operational type of holiday and tourism. The design concept of the building comes from the attractive landscape and excellent ecological environment of Deqing, with the greenery theme merged into architectural image and environmental protection theme run through the inside and outside of the building.

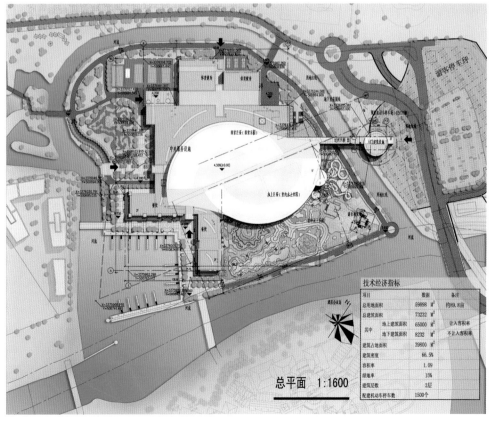

建设单位：开元集团
功能用途：游乐设施
设计/竣工年份：2011年/~
建设地点：浙江省德清县下渚湖
总建筑面积：73 232 m²
总建筑高度/层数：15 m/ 2 F
结构形式：钢混结构
合作单位：加拿大FORREC

Owner: New Century Group
Function: Entertainment Facility
Design/Completion Year: 2011/~
Construction Site: Xiazhu Lake, Deqing County, Zhejiang
Total Floor Area: 73,232 m²
Total Height/Floor: 15 m/ 2 F
Structure: Reinforced Concrete Structure
Cooperative Unit: Canada FORREC Ltd.

萧山浪琴湾（公建部分）
Xiaoshan Langqin Bay (Public Building Part)

建设单位：杭州盈海汽车科技有限公司
功能用途：酒店、商业、办公
设计/竣工年份：2011年/2015年
建设地点：浙江省杭州市萧山区江东工业园区
总建筑面积：89 708 m²
总建筑高度/层数：110 m/18 F
结构形式：框筒结构

Owner: Hangzhou Yinghai Auto Science and Technology Co.,Ltd.
Function: Hotel, Commerce, Office
Design/Completion Year: 2011/2015
Construction Site: Jiangdong Industrial Park, Xiaoshan District, Hangzhou City, Zhejiang
Total Floor Area: 89,708m²
Total Height/Floor: 110 m/18 F
Structure: Frame-Tubel Structure

　　项目位于江东工业园区青六路西侧、江东二路南。总用地面积 33 334 m²。设计力求创造能体现江东发展精神的城市地标性建筑。根据对基地的分析，把地块中两栋主楼前后交错设置，使建筑位于最醒目位置，大大提升了江东工业园区入口的大门形象，同时能更好地共享城市景观资源。主体建筑群落后退形成一个大型的入口广场，作为南面城市景观广场的延续，更加突出了项目的景观优势。建筑造型采用了更加具有人文气息的新古典主义风格，立面色彩也选用了相对温馨的暖黄色调。裙房造型与主楼协调统一，通过连廊连接北侧商业用房，形成一个完整的建筑群落。

This project is located in Jiangdong Industrial Park, on the west side of Qingliu Road and south of Jiangdong Second Road. The total area is 33,334 m². The design tries to create a city landmark that can embody the spirit of Jiangdong's development. According to the analysis of the base, the two main buildings in the center of the area are interleaved to make the buildings in the most prominent position, which would improve the main gate image of Jiangdong Industrial Park and share the city landscape resources at the same time. The backward of the main buildings forms a grand entrance square; as an extension of the city landscape in the south, it highlights the landscape advantage of the project. The buildings apply a neo-classical style with more humanistic atmosphere and the facade color is warm yellow which is relatively cozy. The style of the podium is in accordance with the main building. Connecting with the business buildings in the north through corridor, it forms a complete building community.

杭州恒兴大厦
Hangzhou Hengxing Building

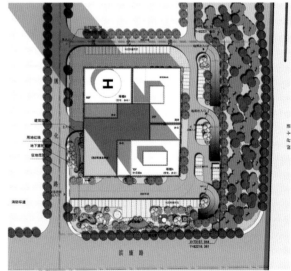

建设单位：杭州恒兴置业有限公司
功能用途：酒店、办公、住宅、商业
设计/竣工年份：2010年/~
建设地点：浙江省杭州市滨江区
总建筑面积：212 700 m²
总建筑高度/层数：239.5 m/70 F
结构形式：框筒结构
合作单位：中建国际（深圳）设计顾问有限公司、Archilier Architecture

Owner: Hangzhou Hengxing Real Estate Co., Ltd.
Function: Hotel, Office, Residence, Commerce
Design/Completion Year: 2010/~
Construction Site: Binjiang District, Hangzhou City, Zhejiang
Total Floor Area: 212,700 m²
Total Height/Floor: 239.5 m / 70 F
Structure: Frame-Tube Structure
Cooperative Unit: China Construction (Shenzhen) Design International Co., Ltd. / Archilier Architecture

项目以239.5 m的高度成为钱塘江南岸的制高点，涵盖了五星级酒店、办公、住宅及商业等多项功能，致力于成为区域标志性建筑物。建筑用地面积偏小，为提升整体建筑效果，建筑裙房底层基本架空，将景观、人流引入建筑内部并提供全天候的公共交流空间，同时解决交通、绿化及消防疏散问题。建筑风格以现代、简洁、时尚为主基调，通过虚实对比区分建筑体量，强调建筑整体的竖向感，并利用竖向构件的变化将卓越集团的菱形标识融入建筑形象之中。

This project becomes the peak of the south bank of Qiantang River with a height of 239.5 m. It covers many functions such as five-star hotel, office, residence, business, etc. It aims to become the landmark of this area. The covering area is smaller. To improve the overall building effect, the bottom of the podium is basically built on stilts. It brings the landscape and people into the building and provides public communicating space 24 hours. At the same time, it solves the problem of traffic, afforesting, and fire evacuation. The style of the building is based on the modern, simple, and fashionable. It differentiates the building volume by virtual-real comparison. It stresses on the vertical sense of the buildings and uses the changes of vertical components to integrate the company's diamond logo to the building.

丽晶国际中心
Regent International Centre

项目位于钱江世纪城核心地块，西临钱江二路。建筑平面布局主要是处理建筑功能形体与场地特征之间的关系，以求达到空间与功能的平衡。整座主体建筑呈S形曲线平面，由此而自然生成南北面独立的景观广场与景观绿化。基地主入口结合北广场水景设置在公园东路。在钱江二路基地西南角增设消防车入口，在建筑四周设环形消防通道。建筑立面外部轮廓简洁、精练，充满时代特征。建筑用材以金属板、玻璃、金属漆、花岗岩等材料协调搭配，塑造建筑恒久特质。

The project is located in the central area of Qianjiang Century CBD and is adjacent to Qianjiang No.2 Road at west. The main purpose of plane building layout is to coordinate the relation between building functions and site features, so as to reach the balance between the space and function. Main building body is a S-shaped curved plane, which naturally creates the independent landscape square and landscape belt on both the south and north sides. Main entrance to the site and the waterscape on the north square are planned on Gongyuan East Road. Fire entrance is planned at southwest corner of the site on Qianjiang No.2 Road, and fire passage is designed around the building. External profile of building facade is simple and compact, embodying modern building features. Metal plate, glass, metallic paint and granite are used to infuse a kind of eternal feature into the building.

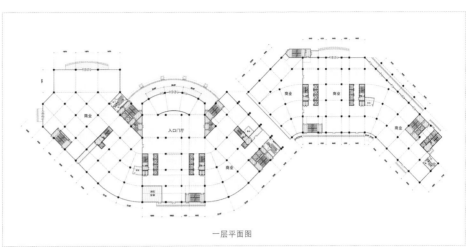

一层平面图

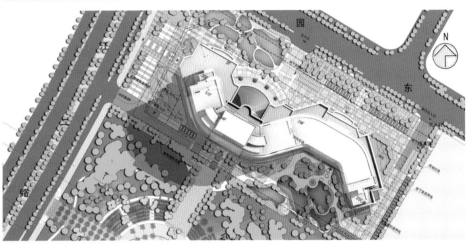

建设单位：浙江尊峪置业有限公司	Owner: Zhejiang Zunyu Real Estate Co., Ltd.
功能用途：商业、办公	Function: Commerce, Office
设计/竣工年份：2011年/2015年	Design/Completion Year: 2011/2015
建设地点：浙江省杭州市钱江世纪城	Construction Site: Qianjiang Century CBD, Hangzhou City, Zhejiang
总建筑面积：280 374 m²	Total Floor Area: 280,374 m²
总建筑高度/层数：194 m/34 F	Total Height/Floor: 194 m / 34 F
结构形式：框筒结构	Structure: Frame-Tube Structure

慈溪财富中心
Cixi Fortune Centre

建 设 单 位：慈溪市工业品批发市场有限公司
功 能 用 途：住宅、商业
设计/竣工年份：2009年/2011年
建 设 地 点：浙江省慈溪市
总 建 筑 面 积：197 493 m²
总建筑高度/层数：99.6 m/30 F
结 构 形 式：框剪结构
合 作 单 位：中国建筑设计研究院

Owner: Cixi Industrial Products Wholesale Market Co., Ltd.
Function: Residence, Commerce
Design/Completion Year: 2009/2011
Construction Site: Cixi City, Zhejiang
Total Floor Area: 197,493m²
Total Height/Floor: 99.6 m/ 30 F
Structure: Frame-Shear Wall Structure
Cooperation Unit: China Architecture Design & Research Group

项目位于慈溪商都二期南侧，地上三十层。地下三层为停车场和机房，地上一层至三层主要为百货商业、超市和银行；地上四层、五层为影院和美食城。六层至三十层为住宅和公寓部分，其中A栋为单身公寓，B、C、D、E栋为住宅。五层屋顶布置有住宅入户公共花园和休闲会所，为住户创造了一个安全、宁静、高雅的活动空间。外立面以深色石材幕墙为主，住宅外立面采用公建风格，和底部商业建筑采用一体化设计手法，整体建筑简洁大气，气势恢宏。

This project is located at the southern part of Cixi Shangdu Phase II. It consists of 30 overground floors and 3 underground floors. The 3-storey basement are used as parking lot and machine room; the bottom 3 ground floors are retail space, supermarket and bank; 4F and 5F are cinema and restaurant; 6F-30F are residential units and apartment, with single apartment in building A and residential units in buildings B, C, D and E. The top five floors are used as entrance garden and leisure chamber, providing a safe, comfortable and elegant public space. The exterior facade is mainly made of dark stone curtain wall. The facade of residential building has a public building appearance and is integrated with the bottom commercial floors, creating a simple but generous overall appearance.

迪安诊断产业基地建设项目
Di'an Diagnostics Industrial Base

建 设 单 位：迪安诊断
功 能 用 途：办公、实验室
设计/竣工年份：2011年/~
建 设 地 点：浙江省杭州市
总 建 筑 面 积：42 226 m²
总建筑高度/层数：30 m/7 F
结 构 形 式：框架结构

Owner: Di'an Diagnostics Co., Ltd.
Function: Office, Laboratory
Design/Completion Year: 2011/~
Construction Site: Hangzhou City, Zhejiang
Total Floor Area: 42,226 m²
Total Height/Floor: 30 m/ 7 F
Structure: Frame Structure

项目位于杭州市西湖区，建设用地面积1.67公顷。基地是一个梯形的用地，分为实验区和行政办公区。实验区主要包括GSP货物仓库及生产经营用房。生产经营用房具体又分为病理科、检验科、科研中心、特检及GMP厂房。行政办公区主要包括办公区、会议区及配套辅助用房。建筑造型风格现代，简洁大气。建筑立面主要由石材和玻璃构成。实验楼与行政办公楼方正纯净，体形硬朗、挺拔，大大加强了建筑的雕塑感。

The project, located in Xihu District of Hangzhou City, covers a site area of 1.67 hm². The trapezoidal site is divided into experimental zone and administrative zone. The experimental zone mainly provides GSP warehouse, production and operation facilities, which are subdivided into pathology department, inspection department, scientific research center, special inspection and GMP factory building, etc. Administrative zone mainly includes working block, conference zone and supporting facilities. The building has a modern, simple and generous appearance. and its facades are mainly decorated with stone and glass. The sculptural laboratory building and the administrative building are regular and pure, embodying strong and straight image.

慈溪市客运中心站新建工程
Cixi Passenger Transportation Center Station

项目位于慈溪市逍林镇，占地面积189 333.3 m²。根据客运中心站的功能性质和使用途径，设计将基地大体分为五个区块：站前广场区、客运中心区、综合服务区、管理办公区及停车场区。站前广场区由广场、公交车、出租车、社会车辆上下客区等组成。客运中心区由进站厅、候车厅、发车位等组成，是旅客上下车的主要活动空间。综合服务区由商场、餐饮、托运、站内办公区以及站内辅助用房区等组成。管理办公区由内部管理用房组成。停车场区由站内车辆停车区及外来车辆停车区组成。

汽车客运中心站建成后将成为集中慈溪市长途、城市公交、城乡公交、出租车、轨道交通和货物零担等在内的一级客运站综合交通枢纽。

The project is located in Xiaolin Town, Cixi City, covering a site area of about 189,333.3 m². Based on functional features and purposes of passenger transportation center station, the design divides the whole site into five zones, including station square, passenger center, general service zone, management zone and parking lot. The station square is composed of square area and loading/unloading areas for bus, taxi and social vehicle, etc. Passenger center includes checking hall, waiting room and departure area, which are main space for passengers to get on/off. General service zone provides retail space, restaurant, luggage check-in, station office and station auxiliary room, etc. Management zone is composed of internal management rooms. The parking lot consists of areas for internal vehicles and for exotic vehicles.

After completion, this passenger transportation center will become grade-I general passenger transportation hub in Cixi, integrating various transportation functions, such as long-distance coach, urban bus, rural bus, taxi, rail transit and goods freight, etc.

建 设 单 位：慈溪市交通局
功 能 用 途：交通
设计/竣工年份：2011年/～
建 设 地 点：浙江省慈溪市
总 建 筑 面 积：79 600 m²
总建筑高度/层数：80 m /22 F
结 构 形 式：框架结构、网架结构

Owner: Cixi Traffic Bureau
Function: Traffic
Design/Completion Year: 2011/~
Construction Site: Cixi City, Zhejiang
Total Floor Area: 79,600 m²
Total Height/Floor: 80 m/22 F
Structure: Frame Structure, Space Truss Structure

乌什县社会综合福利中心
Wushi Social General Welfare Center

项目从功能上分成六个部分，设计结合建筑形体的塑造和空间的布局，将儿童福利院布置于1#楼西楼部分，共三层，一层布置大教室、活动室、办公室；二、三层布置六人间儿童宿舍。

福利中心的外观造型，以入口弧形连廊围合的半圆形广场为核心展开，采用涂料和窄窗的外墙立面，以体现稳重、大方的建筑基调，同时局部的预制花格墙与墙面形成对比，体现出现代建筑气息。建筑整体雕塑感强烈，展现出一个既具有地标性，又稳重开放、具有亲和力的新型社会福利中心的建筑形象。

This project is divided into six functional zones, and the design scheme focuses on building shape and space layout to plan the children welfare institute in the 3-storey west volume of building 1#, where 1F is used as large classroom, activity room and office, and 2F and 3F are used as 6-bed children dormitory. The building shape of welfare center unfolds around the semicircular square enclosed by the entrance arched corridor. Its facades are decorated with coating and narrow window openings, to express its solemn and elegant building image. The intensive contrast between the local prefabricated lattice wall and solid wall highlights modern building features, and the overall sculpture image of the building realizes a landmark building of new social welfare center boasting of solemn, opening and welcoming appearance.

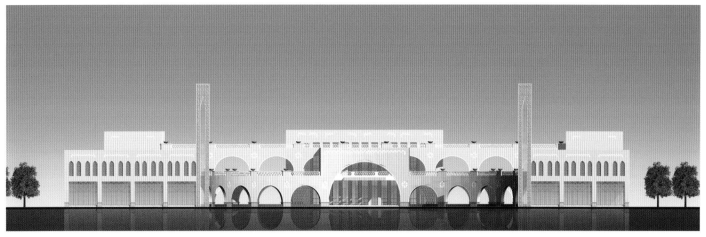

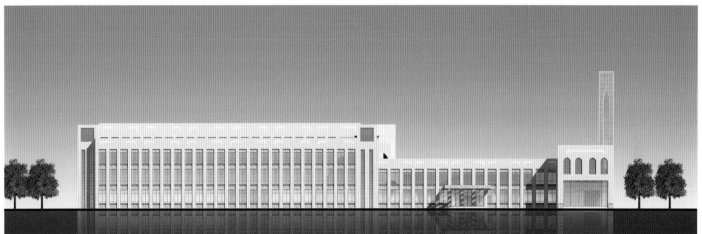

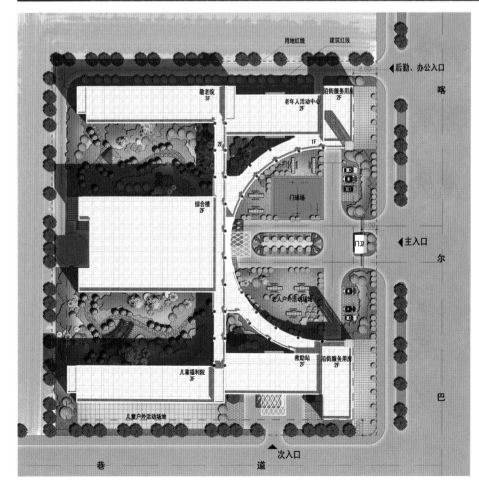

建设单位：浙江衢州市援疆建设指挥部
功能用途：公共服务
设计/竣工年份：2011年/2013年
建设地点：新疆乌什县
总建筑面积：10 380 m²
总建筑高度/层数：13.8 m /3 F
结构形式：框架结构

Owner: Quzhou Xinjiang Assistance and Construction Office
Function: Public Service
Design/Completion Year: 2011/2013
Construction Site: Wushi County, Xinjiang
Total Floor Area: 10,380 m²
Total Height/Floor: 13.8 m /3 F
Structure: Frame Structure

浙江国际影视中心综合大楼
Zhejiang International Film and TV Center General Building

建 设 单 位：浙江广播电视集团
功 能 用 途：影视制作、演播、办公
设计/竣工年份：2011年/2015年
建 设 地 点：浙江省杭州市萧山区
总 建 筑 面 积：190 243 m²
总建筑高度/层数：218 m /42 F
结 构 形 式：框筒结构

Owner: Zhejiang Broadcasting and Television Group
Function: Film and TV Production, Broadcasting, Office
Design/Completion Year: 2011/2015
Construction Site: Xiaoshan District, Hangzhou City, Zhejiang
Total Floor Area: 190,243 m²
Total Height/Floor: 218 m/42 F
Structure: Frame-Tube Structure

浙江国际影视中心被列入浙江省"十一五"重大建设项目和浙江省文化产业"四个一批"项目。规划面积约282 667 m²，是一座集影视制作、演播、创意、影视培训及相关配套设施于一体的文化产业园。

影视后期制作综合大楼是影视中心的主要建筑，建筑面积约200 000 m²，裙房部分包括8个大小不同的演播厅，地上塔楼为42层。建筑立面以简洁明快的玻璃幕墙系统，配合弥漫着金属光泽的金属铝板以及色彩协调的石材幕墙，既大气端庄，又彰显着高新科技的成就，体现了现代的时尚美感。

Zhejiang International Film and TV Center is listed as one of "The 11th Five-Year Plan" key construction projects and cultural industrial "The Four First Batches" of Zhejiang Province. It covers a site area of about 282,667 m² and is a cultural industrial park integrating various functions, such as film and TV production, broadcasting, creative idea, film and TV training and related supporting facilities, etc.

Film and TV post production general building is a main building in this film and TV center, covering a building area of about 200,000 m². It is composed of a podium building including 8 performance and broadcasting halls of different sizes and 42-storey upper tower. Building facades are decorated with simple and bright glass curtain wall system, shining aluminium panel and colorful stone curtain wall, to embody elegant, solemn and high-tech features, as well as modern aesthetics.

香山四季中心
Xiangshan Season Center

项目位于杭州市萧山区亚太路以南，绕城高速沿线以西，占地面积为15 059 m²。本工程为办公和商业性质的3栋高层建筑和两栋低层建筑。高层建筑层数为地上19层（1层裙房），地下2层，属一类高层建筑，建筑的耐火等级为一级。1、2号楼高度为99.82 m；3号楼82.4 m；4、5号楼9.815 m。裙房底层架空，布置绿化、办公、商贸及门厅。1、2号楼3至18层为商贸用房，9至19层为办公用房；3号楼为办公用房；4、5号楼为商贸用房。

The project is located at south of Yatai Road in Xiaoshan District, Hangzhou, and west of urban ring high-way, covering a total area of 15,059 m². This project is composed of three high-rise buildings and two low-rise buildings, providing offices and retail spaces. The high-rise buildings are class-I high-rise buildings with grade-I fire resistance, each containing 19-storey overground volume (1-storey podium building) and 2-storey basement. The height is 99.82 m for buildings 1 and 2, 82.4 m for building 3 and 9.815 m for buildings 4 and 5. Ground floor of the podium building is supported in the air to act as an entrance hall for green area, offices and retail spaces. 3F-18F of buildings 1 and 2 are retail spaces, 9F-19F are offices; building 3 provides offices; buildings 4 and 5 are used as retail spaces.

建 设 单 位：浙江鼎盛实业发展有限公司	Owner: Zhejiang Dingsheng Real Estate Development Co., Ltd.
功 能 用 途：商业、办公	Function: Commerce, Office
设计/竣工年份：2010年/2013年	Design/Completion Year: 2010/2013
建 设 地 点：浙江省杭州市萧山区	Construction Site: Xiaoshan District, Hangzhou City, Zhejiang
总 建 筑 面 积：102 121 m²	Total Floor Area: 102,121 m²
总建筑高度/层数：99.8 m/19 F	Total Height/Floor: 99.8 m/19 F
结 构 形 式：框剪结构	Structure: Frame-Shear Wall Structure

浦江县广播电视中心迁建项目一期工程
Pujiang County Broadcasting and Television Center Relocation Project Phase I

建设单位：浦江广播电视台
功能用途：办公、广播中心
设计/竣工年份：2010年/2013年
建设地点：浙江省浦江县
总建筑面积：20 689 m²
总建筑高度/层数：49.95 m/12 F
结构形式：框架结构

Owner: Pujiang Broadcasting and Television Station
Function: Office, Broadcasting Center
Design/Completion Year: 2010/2013
Construction Site: Pujiang County, Zhejiang
Total Floor Area: 20,689 m²
Total Height/Floor: 49.95 m/ 12 F
Structure: Frame Structure

项目位于平七路中埂段以东。在建筑布局上，我们通过最基本的几何图形语言，在网格式结构的基础中寻求变化，有秩序地布置建筑、广场与水景绿化。将12层主楼设于基地北侧，坐北朝南。裙楼靠地块东侧展开80 m。立面造型上我们在主楼和裙楼上均采用竖向线条，与东面档案馆的立面元素相协调，铝合金竖线条直接落地，体现建筑的简洁、庄重与挺拔感。裙楼的竖线条与主楼立面呼应，将整组建筑塑造成一个有着鲜明特色的新时代广电建筑。

The project is located at east of Zhonggeng Section of Pingqi Road. The building layout uses the most fundamental geometrical language to create diversification in lattice structure foundation and to realize regular arrangement of buildings, square and waterscape elements. The 12-storey main building stands on the north part of the site and faces to the south. Podium building is adjacent to the east border of the site and stretches 80 meters long. The facades of main building and podium building are decorated with longitudinal lines to respond to the facade elements on archives at east. The longitudinal aluminium alloy strips stretch from floor to roof, highlighting the simple, solemn and straight appearance of the building. Longitudinal lines on facade of podium building respond to the facade of main building, creating a modern broadcasting and television building with vivid image.

杭州临平永安·金鑫大厦
Hangzhou Linping Yong'an · Jinxin Mansion

建设单位：杭州永安房地产开发有限公司
　　　　　杭州金鑫外贸有限公司
功能用途：办公、商业
设计/竣工年份：2010年/2013年
建设地点：浙江省杭州市余杭区迎宾大道
总建筑面积：80 000 m²
总建筑高度/层数：100 m/26 F
结构形式：框筒结构

Owner: Hangzhou Yongan Real Estate Development Co., Ltd.
　　　　Hangzhou Jinxin Foreign Trade Co., Ltd.
Function: Office, Commerce
Design/Completion Year: 2010/2013
Construction Site: Yingbin Boulevard, Yuhang District, Hangzhou City, Zhejiang
Total Floor Area: 80,000 m²
Total Height/Floor: 100 m/ 26 F
Structure: Frame-Tube Structure

　　项目位于杭州市余杭区临平镇迎宾大道东侧，西侧紧临地铁终点站。本案两块用地相互毗邻，将两个地块合二为一进行设计，这样既使得建筑造型大气统一，空间丰富灵动，又使得资源优势得到最大化的价值体现。双塔的南面设有一个大型开敞的景观广场，西南面亦可享受近在咫尺的城市绿地。方形对称的建筑体形，一方面使建筑形体在每个城市界面显得修长挺拔，同时也创造了一个在各个城市角度均具有标志性形象的建筑。

The project is located at east of Yingbin Boulevard of Linping Town in Yuhang District of Hangzhou and is adjacent to a subway terminal at west. It is composed of two adjacent plots which are designed as a whole. This results in uniform and generous building shape, diversified and flexible spaces, and best expression of advantageous resources. The two towers face to a large and open landscape square at the south and enjoy an urban green land at the southwest. Symmetrical square building shape creates upsoaring and straight skyline, and creates a landmark building image.

慈溪文化商务区03地块
Cixi Cultural Business District Plot 03

项目地处慈溪市文化商务区，新城大道东侧，是一组以文化建筑为核心，以办公、商业、居住复合型城市功能为基础的新型滨水城市空间。建筑立面的造型采用简洁的现代风格，四幢塔楼高低错落，占据地块的四角。建筑立面采用全幕墙设计，高耸的塔体在上部作斜切处理，形成多个斜切面——A楼的东北角、西南角，B楼的东南角、西北角，C楼的东南角、西北角，D楼的东北角、西南角，通体透亮的玻璃幕墙与斜切面相结合，如钻石般熠熠生辉，将现代城市的高效与多样诠释得淋漓尽致。

The project is located in the cultural business district of Cixi City and on the east side of Xincheng Boulevard. It is a group of new waterfront urban spaces focusing on cultural buildings and based on composite urban functions, such as office, retail and residence, etc. The buildings adopt simple and modern facades and four tower buildings standing on four corners of the site respectively have different heights. Building facades are fully covered by curtain wall and the tops of tower buildings are chamfered to form multiple diagonal planes, including northeast and southwest corners of Building A, southeast and northwest corners of Building B, southeast and northwest corners of Building C, northeast and southwest corners of Building D. Transparent glass curtain wall and the diagonal planes form diamond building shape, clearly interpreting the high efficiency and diversified features of modern urban space.

建设单位：浙江承兴/慈溪世茂/宁波美华/宁波凯玛	Owner: Zhejiang Chengxing/Cixi Shimao/Ningbo Meihua/Ningbo Kaima
功能用途：商业、办公	Function: Commerce, Office
设计/竣工年份：2010年/2013年	Design/Completion Year: 2010/2013
建设地点：浙江省慈溪市	Construction Site: Cixi City, Zhejiang
总建筑面积：181 126 m²	Total Floor Area: 181,126m²
总建筑高度/层数：99.9 m/26 F	Total Height/Floor: 99.9 m/ 26 F
结构形式：框剪结构	Structure: Frame-Shear Wall Structure

乐清总部经济园
Yueqing Headquarters Economic Park

项目位于乐清中心区东南端，交通便捷，环境优美。建筑造型以地块东北角的超高层建筑为起点，通过形体间不同的连接方式将各个单体串联成一个整体，高低起伏，错落有致，恰似一条正欲起飞的巨龙。尤其是沿旭阳路的建筑，形体上下转承更为生动，恰如"龙腾四海"，象征着企业蒸蒸日上，兴旺发达。

The project is located at the southeast end of the Central District of Yueqing, enjoying convenient traffic and beautiful environment. Starting from the super high-rise building standing at the northeast corner of the plot, the independent buildings are connected through different connection methods into an integral entity of diversified elevations, just like a soaring dragon and the buildings on Xuyang Road has a vivid image, embodying the booming development of the enterprise.

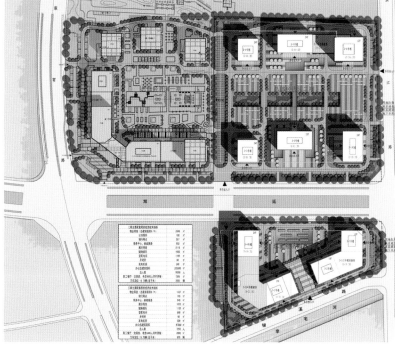

建 设 单 位：乐清市中心区发展有限公司	Owner: Yueqing Central District Development Co., Ltd.
功 能 用 途：办公	Function: Office
设计/竣工年份：2010年/2012年	Design/Completion Year: 2010/2012
建 设 地 点：浙江省乐清市中心区	Construction Site: Central District, Yueqing City, Zhejiang
总 建 筑 面 积：869 000 m²	Total Floor Area: 869,000 m²
总建筑高度/层数：194.2 m/47 F	Total Height/Floor: 194.2 m/ 47 F
结 构 形 式：框筒、钢管、钢混结构	Structure: Frame-Tube Structure, Steel Tubular Structure, Reinforced Concrete Structure

天津德力置业
Tianjin Deli Real Estate

建 设 单 位：天津德力置业有限公司
功 能 用 途：综合
设计/竣工年份：2009年/～
建 设 地 点：天津生态城
总 建 筑 面 积：30 760 m²
总建筑高度/层数：23.3 m/6 F
结 构 形 式：框架结构

Owner: Tianjin Deli Real Estate Co., Ltd.
Function: Complex
Design/Completion Year: 2009/~
Construction Site: Ecological City, Tianjin
Total Floor Area: 30,760 m²
Total Height/Floor: 23.3 m/ 6 F
Structure: Frame Structure

设计根据地形条件布置建筑及道路广场。商业中心的主要入口在东南角，酒店和办公的入口分别位于西南角，餐饮入口位于北侧，东侧为商业人行入口。在地块的东北角和东南角各设一个机动车出入口，车流入口与人流入口完全分开。三个主体建筑围合成一个院式的内广场，同时设置三个出入口进入内广场。地块北面建筑主要为餐饮用房，一至三层适当布置商铺，其余建筑一至三层为商业用房，L形建筑西侧四层以上布置办公用房，南侧四层以上布置酒店用房。地下室北面布置超市，其余为机动车停车库。

Based on the terrain conditions, the buildings, roads and squares are arranged. The main entrance of the commercial center is in the southeast corner. The entrances of the hotel and office buildings are in the southwest corner, the catering entrance is at the north, and the commercial pedestrian entrance is at the east. There is an access for vehicles in the northeast and southeast corner. The streams of traffic and people are separate. The three main buildings form a courtyard-style internal square and there will be three entrances for entering the internal square. The buildings in the north are mainly for catering. Shops are arranged from the ground floor to the second floor. The floors from the ground floor to the second floor are commercial spaces in the rest of the building. The floors above the third-floor of the L-shaped building in the west are office buildings and hotel buildings for that of the floors in the south. A supermarket is located at the northern part of the basement and the rest is the parking garage for vehicles.

常州亚细亚影城
Asia Film Center in Changzhou

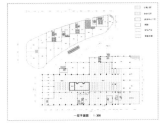

建设单位：常州广电集团
功能用途：商业、办公
设计/竣工年份：2009年/~
建设地点：江苏省常州市
总建筑面积：68 000 m²
总建筑高度/层数：80 m/31 F
结构形式：框剪结构

Owner: Changzhou Broadcasting and TV Group
Function: Commerce, Office
Design/Completion Year: 2009/~
Construction Site: Changzhou City, Jiangsu
Total Floor Area: 68,000 m²
Total Height/Floor: 80 m/ 31 F
Structure: Frame-Shear Wall Structure

亚细亚影城有着不凡的历史与影响，地块有着特殊的地理区位。设计中运用大量柔美的曲线作为母题来刻画建筑。建筑主体犹如一条微微抬头的巨龙，同闪耀的群星，向无尽的星空缓缓伸展，塑造了地块的整体形象。在新建建筑和原有的建筑之间我们精心设计了一条高达20m的气势恢宏的商业街，这条商业街内组织了整个建筑群主要的商业娱乐空间。步行街对原有建筑和新建建筑进行了无缝连接，同时通过对原亚细亚影城的形象统一改造，使完成后的建筑浑然一体，犹如一个全新统一建设的建筑群，具有一个整体的令人难以忘记的形象。

The Asia Film Center has a remarkable history and impact and has special geographical location. A large amount of soft curves are applied in the design as the theme to characterize the building. The main building is like a dragon that slightly raises its head, extending to the endless starry sky with shining stars, which makes the overall image of the area. We have designed a grand commercial street between the new and old buildings that is up to 20 meters in height. The main commercial and entertainment space is organized in this street. The pedestrian street seamlessly connected the old buildings and the new buildings. At the same time, the image of original Asia Film Center is transformed and it makes the completed buildings become one, which is like new unified buildings and has an overall impressive image.

杭州萧山国际机场T3航站楼
Xiaoshan International Airport T3 Terminal Building, Hangzhou

萧山国际机场二期扩建工程分两个阶段实施，第一阶段的主体工程国际航站楼已于2010年7月完工并投入使用，T3航站楼则是第二阶段的主体工程，它与T1航站楼无缝衔接。将来通过对T1航站楼前厅屋面及幕墙的改造，使T3和T1航站楼成为一个统一的整体，航站楼主体的波浪形屋面从两侧由低到高逐渐上升，形成新的中轴对称的富有动感的建筑造型，不仅延续了原有建筑的波浪概念，而且还体现出新的江南水文化的风格。T3航站楼内部设计坚持以人为本的原则，平面功能合理、旅客流程顺畅，营造出怡人的空间环境，并且取得了设计阶段的绿色建筑"三星级"。

Xiaoshan International Airport Phase II expansion project is constructed in two phases. Main content of the first phase is an international terminal building which has been completed and put into use in July of 2010. T3 terminal building is the main content of the second phase and it will realize seamless connection with T1 terminal building. Lobby roof and curtain wall of T1 terminal building will be rebuilt in the future to form an integral volume with T3 terminal building. The main wave roof of the terminal building gradually goes up from both sides and forms a new dynamic building image symmetrically along the axis, not only inheriting the wave shape of original building, but also embodying the water cultural style in South Yangtze River Region. Interior design of T3 terminal building follows the human-oriented principle to realize a reasonable plane functional layout, smooth passenger communication, and to create welcoming and comfortable space environment. This project has been evaluated as 3-star green building in its design stage.

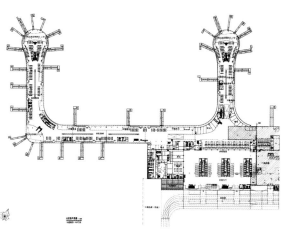

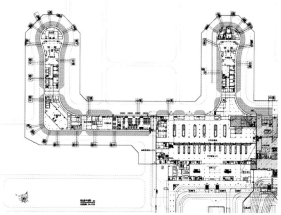

建设单位：萧山国际机场二期工程建设指挥部
功能用途：交通
设计/竣工年份：2009年/2012年
建设地点：浙江省杭州市萧山机场
总建筑面积：170 534 m²
总建筑高度/层数：31.4 m/4 F
结构形式：框架结构、钢结构
合作设计：华东建筑设计研究院有限公司

Owner: Xiaoshan International Airport Phase II Construction Office
Function: Traffic
Design/Completion Year: 2009/2012
Construction Site: Xiaoshan International Airport, Hangzhou City, Zhejiang
Total Floor Area: 170,534 m²
Total Height/Floor: 31.4 m/4 F
Structure: Frame Structure, Steel Structure
Cooperation Unit: East China Architectural Design & Research Institute Co.,Ltd.

华能长兴电厂去工业化设计
Huaneng Changxing Power Plant De-industrialization Design

建设单位：华能长兴电厂
功能用途：工厂
设计/竣工年份：2012年/~
建设地点：浙江省湖州市长兴县
总建筑高度：240 m
结构形式：钢结构
合作单位：浙江电力设计院

Owner: Huaneng Changxing Power Plant
Function: Plant Factory
Design/Completion Year: 2012/~
Construction Site: Changxing County, Huzhou City, Zhejiang
Total Height: 240 m
Structure: Steel Structure
Cooperation Unit: Zhejiang Electric Power Design Institute

项目位于湖州市长兴县吕山乡。

设计从规划到单体进行各个层面的控制，从厂址所在环境入手，深入挖掘设计的出发点，结合项目的总体布局、总平面布置、建筑物的外观、厂区周边环境，打造与当地生态园林城市相融合的去工业化设计方案，力求使电厂在满足工业要求的基础上，具有更强的艺术性。

我们在设计中用竹子作为烟囱的形象，用丝绸作为冷却塔的外表皮素材，用山水剪影作为锅炉房与主厂房的造型示意，追求新元素的大胆运用，使整个工业厂区摒弃往常冰冷乏味的感觉，取而代之的是活泼、鲜明、具有时代特色的建筑风格。

This project is located in Lvshan Village, Changxing Town, Huzhou City.
The design controls all levels from the planning to the unit. It starts from the environment at the site and digs out the starting point of the design. Combining the overall layout, general layout, appearance of the buildings, and the surroundings of the plant, it creates the de-industrialization plan that integrates with local ecological garden and tries to make the plant more artistic on the basis of meeting the technological requirements. We apply the bamboo image as a chimney in the design, silk as the external skin of the cooling tower, and landscape as the modeling indication of boiler room and main plants. We seek for the bold application of new elements to make the whole industrial area abandon the traditional cold and boring feeling; instead, there is a lively, distinctive and fashionable architectural style.

杭州热电厂地块（局部）概念性方案设计
Hangzhou Thermal Power Plant (Partial) Concept Plan

建设单位：杭州城投集团
功能用途：城市综合体
设计/竣工年份：2009年/~
建设地点：浙江省杭州市北大桥
总建筑面积：107 912 m²
总建筑高度/层数：150 m / 40 F
结构形式：砖混、框架结构

Owner: Hangzhou Urban Construction and Investment Group
Function: Urban Complex
Design/Completion Year: 2009/~
Construction Site: Beidaqiao, Hangzhou City, Zhejiang
Total Floor Area: 107,912 m²
Total Height/Floor: 150 m / 40 F
Structure: Masonry-Concrete Structure, Frame Structure

　　项目位于北大桥化工区，用地面积约38 hm²。本规划区定位是集休闲旅游、商务办公、文化娱乐于一体，兼具功能齐全的高品质公共服务的旅游综合体以及高品质居住社区的多功能综合区域。
　　在设计中以"应保尽保、积极保护"为原则，通过局部立面整治和内部装修，形成工业遗产保护区块兼创意产业园。同时以烟囱为中心，结合下沉广场、树阵、旱喷和热电设备室外展示区，打造出一个休憩及工业设备展示广场。

This project is located in the Beidaqiao chemical engineering area and covers an area of 38 hm². The positioning of this design is: integrating leisure and tourism, business and office, and culture and entertainment, to create a tourism complex with advanced public service of complete functions, as well as a multi-functional comprehensive area with high-quality residence.
The design is based on the rule of "try best to and actively protect the existing things". The industrial heritage and creative industrial park will be formed through partial renovation and internal decoration. At the same time, the chimney is the center and with the sunken square, trees, dry spray, and thermal power equipment as the external display area a display square for recreation and industrial equipment will be created.

办公建筑
Office Building

62	钱江世纪城J—01—03地块项目
64	衢州市社会保障服务中心
66	新疆阿拉尔市行政中心办公楼
68	新疆阿拉尔市法院、检察院
70	交通银行股份有限公司宁波分行大厦
71	山东东营市广饶县颐和国际大厦
72	超威电源有限公司新能源大厦
74	衢州国土资源局办公大楼
75	长兴电力调度及生产运检用房
76	湖州市吴兴区民防建设大楼
78	厦门福兴国际中心
79	黄岩区政府综合服务中心
80	杭州丰盛·恒泰广场
82	海正药业（杭州）有限公司总部大楼
83	衢州建设局办公大楼
84	湖州电力生产运维基地
86	世华大厦
87	慈溪龙山镇灵峰路综合写字楼
88	诸暨城东中心区A2地块综合办公大楼
90	正凯中心

62	Qianjiang Century CBD Plot J-01-03
64	Quzhou Social Security Service Center
66	Xinjiang Alaer Administrative Center Office Building
68	Xinjiang Alaer Municipal Court and Procuratorate
70	Bank of Communication Ningbo Branch Mansion
71	Guangrao County Yihe International Mansion, Dongying City, Shandong
72	New Energy Tower of Chilwee Power Co., Ltd.
74	Quzhou State Land Resource Bureau Office Building
75	Changxing Power Bureau Building for Power Dispatching and Production Maintenance
76	Wuxing District Civil Defense Construction Building, Huzhou City
78	Xiamen Fuxing International Center
79	Huangyan District Governmental General Service Center
80	Hangzhou Fengsheng • Hengtai Plaza
82	Headquarters of Hisun Pharmaceutical (Hangzhou) Co., Ltd.
83	Office Building of Quzhou Construction Bureau
84	Huzhou Power Generation and Maintenance Base
86	Shihua Tower
87	General Office Building on Lingfeng Road of Longshan Town, Cixi
88	General Office Building on Plot A2 of Eastern Central District, Zhuji
90	Zhink Center

92	三宏国际大厦
94	杭州和茂大厦
95	麦道大厦
96	台州中央商务区3—02，04地块设计
98	阿里巴巴B2B二号园区
100	杭州汇鑫大厦
102	国金中心
103	杭州华联钱塘会馆
104	温岭九龙商务中心（办证中心）
105	青川县行政中心
106	宁波银行总部大厦
108	钱江新城金融中心
109	杭州日信国际中心
110	潮峰钢构集团有限公司幕墙生产车间改造工程
111	江西德兴市德兴大厦
112	余姚商会大厦
113	杭州协和大厦
114	宁波银行象山大厦
115	杭州经济技术开发区水质检测、调度、控制中心综合楼

92	Sanhong International Building
94	Hangzhou Hemao Building
95	Maidao Mansion
96	Taizhou CBD Plots 3-02 and 04
98	Alibaba B2B No.2 Park
100	Hangzhou Huixin Mansion
102	International Financial Center
103	Hangzhou UDC Qiantang Chamber
104	Wenling Jiulong Business Center (Certificate Service Center)
105	Qingchuan Administration Center
106	Bank of Ningbo Headquarters
108	Qianjiang New Town Financial Center
109	Hangzhou Rixin International Center
110	Curtain Wall Production Plant Reconstruction of Triumpher Steel Structure Group Co., Ltd.
111	Jiangxi Dexing Mansion
112	Yuyao Chamber of Commerce Building
113	Hangzhou Xiehe Building
114	Bank of Ningbo Xiangshan Building
115	Water Quality Inspection, Dispatching and Control Center General Building in Hangzhou Economic and Technological Development Zone

钱江世纪城J—01—03地块项目
Qianjiang Century CBD Plot J-01-03

建 设 单 位：杭州信融置业有限公司
功 能 用 途：商业、办公
设计/竣工年份：2012年/~
建 设 地 点：浙江省杭州市萧山区钱江世纪城
总 建 筑 面 积：104 665 m²
总建筑高度/层数：180 m/42 F
结 构 形 式：框筒结构

Owner: Hangzhou Xinrong Real Estate Co., Ltd.
Function: Commerce, Office
Design/Completion Year: 2012/~
Construction Site: Qianjiang Century CBD, Xiaoshan District, Hangzhou City, Zhejiang
Total Floor Area: 104,665 m²
Total Height/Floor: 180 m/ 42 F
Structure: Frame-Tube Structure

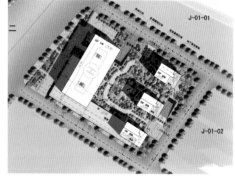

　　项目地块临近地铁出入口、庆春路隧道与建设中的奥体博览中心。项目体现综合性的功能，通过梳理与整合，其静态功能(如服务空间等)与动态功能(如人车动线等)达到合理高效，同时又与建筑的形体相契合。在立面处理上，采用简洁而有力度的形体和中国古建筑中"花格窗"的肌理，以现代的建筑材质进行装饰，从而使现代高层建筑与中国古典装饰艺术完美结合，产生强烈的视觉冲击力，同时也塑造了独特的标志性建筑形象。

The site is adjacent to a subway station, Qingchun Road Tunnel, and Olympic Expo Center which is under construction. It contains static functions, such as service spaces, etc., and dynamic functions, such as pedestrian and vehicle flow, etc., which are reasonably integrated to reach their best efficiency. Meanwhile, functional layout highlights the building's appearance. On the aspect of facade treatment, the "lattice window" texture of Chinese antique building is designed on simple but generous appearance, which is decorated with modern building materials, realizing perfect integration between modern high-rise building and Chinese classical decorative esthetics. This results in intensive visual impact and creates a unique landmark building image.

衢州市社会保障服务中心
Quzhou Social Security Service Center

项目位于浙江省衢州市西区新城中心区内，地块呈矩形，为原衢州市棉纺织厂旧址。设计将建筑主体布置于原棉纺织厂厂房范围内，将高层主楼布置于地块北侧，主要裙房布置于地块中心，辅助裙房布置于地块西南侧无大型乔木的区域，东面保留原始慢行系统和高大乔木，南面建筑退让红线形成大型绿化广场，结合布置建筑的主入口广场，地块的西北角布置大型停车场。在裙房屋顶下布置开放空间以贯通慢行系统，建筑东侧设计次入口广场和入口灰空间。

The project is located on a regular rectangular site in the central area of the new town in west district of Quzhou and this is the location of old Quzhou Cotton Textile Factory. Main buildings are planned on the original factory plant site, with the high-rise main building on the northern part, main podium building on the central part, supporting podium building on the southwestern part where there is no tall arbor. Original walking path and tall arbors are reserved on the eastern part and the building is recessed back from the red line at southern part to form a large planted square, where a main entrance square is designed. A large parking lot is planned on the northwestern part. Open space under the podium building roof is connected with the walking path; the side entrance square and the entrance ash space are designed at east of the building.

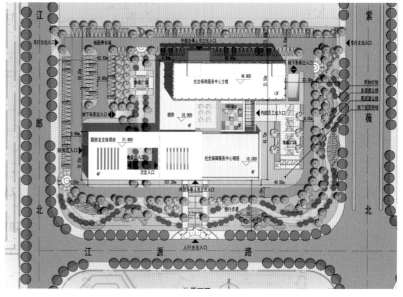

建 设 单 位：衢州市人力资源和社会保障局
功 能 用 途：办公
设计/竣工年份：2013年/~
建 设 地 点：浙江省衢州市
总 建 筑 面 积：34 100 m²
总建筑高度/层数：46.8 m/12 F
结 构 形 式：框架结构

Owner: Quzhou Bureau of Human Resource and Social Security
Function: Office
Design/Completion Year: 2013/~
Construction Site: Quzhou City, Zhejiang
Total Floor Area: 34,100 m²
Total Height/Floor: 46.8 m / 12 F
Structure: Frame Structure

新疆阿拉尔市行政中心办公楼
Xinjiang Alaer Administrative Center Office Building

本项目为改扩建项目，原有主楼处于阿拉尔市三五九文化纪念中心中心轴线之上。新建建筑延续中心对称布局，设置于中心主楼的两侧，自成围合布局，各自形成中心花园。建筑群体通过弧形连廊相互联系，南侧结合景观水系设置大型迎宾广场。在整个总平面布置中，建筑群体呈中心对称，整体布局庄重，大气而恢宏。通过弧形的道路及绿化布置形成优美的迎宾广场，与规整的行政中心建筑群相互融合，使整个基地在表现庄重大气的同时也充满了活力。

This is an expansion and reconstruction project. The original main building is located on the central axis of 359 Cultural Monumental Center of Alaer City. The new building, following the centrosymmetrical layout, is arranged on both sides of the main building to enclose independent central gardens. Buildings are connected through arched overhead corridors and large welcome square is designed to highlight the water landscape at the south. In the overall plan layout, buildings are distributed in a centrosymmetrical layout to express a solemn and generous image. Curved roads and green elements are arranged to form a beautiful welcome square, which is perfectly integrated into the building cluster, infusing great vigor into the whole site.

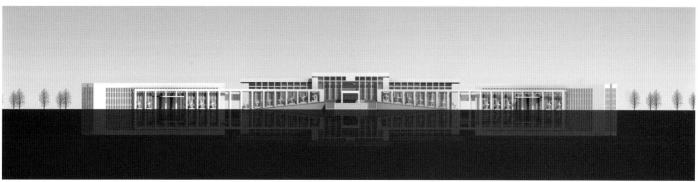

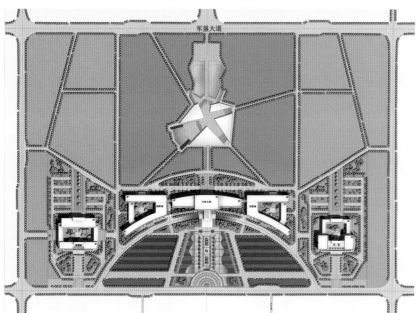

建设单位：新疆阿拉尔市西北兴业城市建设投资发展有限公司
功能用途：综合服务、办公
设计/竣工年份：2012年/~
建设地点：新疆阿拉尔市
总建筑面积：33 733 m²
总建筑高度/层数：13.4 m/3 F
结构形式：框架结构

Owner: Xinjiang Alaer Northwest Xingye Urban Investment and Development Co., Ltd.
Function: General Service, Office
Design/Completion Year: 2012/~
Construction Site: Alaer City, Xinjiang
Total Floor Area: 33,733 m²
Total Height/Floor: 13.4 m/3 F
Structure: Frame Structure

新疆阿拉尔市法院、检察院
Xinjiang Alaer Municipal Court and Procuratorate

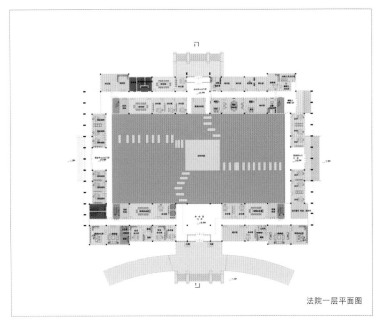

法院一层平面图

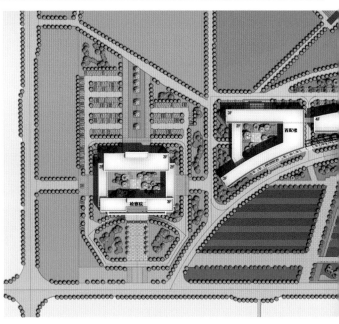

项目基地位于阿拉尔行政中心东侧。法院与检察院建筑对称布置于行政中心两翼。建筑延续中心对称布局，通过"回"字形的建筑布局，围合形成大型中心景观庭院，为使用者提供优质的办公空间。在建筑南侧主入口布置大型柱廊，配合东西两侧布置柱阵及浮雕，营造出国家机器执法机关庄重、大气的场所氛围，呼应行政中心整体建筑形象，融入城市脉络。配合建筑入口布置的小型绿化广场既是法院、检察院独立的入口形象，又能融入行政中心大广场之中，形成西部新兴城市的亮丽风景。

The site is located at east of Alaer Administrative Center. The court and the procuratorate are arranged on different sides of the administrative center. Buildings follow the centrosymmetrical layout and are arranged to form two homocentric circles, which enclose a large central landscape courtyard, providing high-quality office spaces for the users. Large colonnade at the south main entrance cooperates with columns and embossment on the east and west facades to create a solemn and generous location image of national law-enforcement authority. It responds to the overall building image of the administrative center and is integrated into the urban context. Small green square is planned at the entrance to embody independent entrance image of the court and the procuratorate and to merge into the larger square of the administrative center, providing beautiful landscape for this western new boom city.

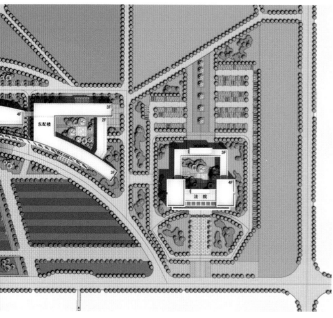

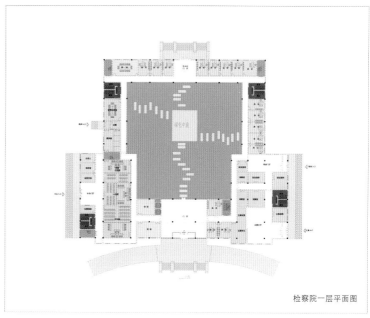

检察院一层平面图

建设单位：新疆阿拉尔市西北兴业城市建设投资发展有限公司
功能用途：综合服务、办公
设计/竣工年份：2012年/~
建设地点：新疆阿拉尔市
总建筑面积：8 748 m²/ 12 322 m²
总建筑高度/层数：12.6 m/3 F；15.6 m/4 F
结构形式：框架结构

Owner: Xinjiang Alaer Northwest Xingye Urban Investment and Development Co., Ltd.
Function: General Service, Office
Design/Completion Year: 2012/~
Construction Site: Alaer City, Xinjiang
Total Floor Area: 8,748 m² / 12,322 m²
Total Height/Floor: 12.6 m/ 3 F；15.6 m/4 F
Structure: Frame Structure

交通银行股份有限公司宁波分行大厦
Bank of Communication Ningbo Branch Mansion

建 设 单 位：交通银行股份有限公司
功 能 用 途：办公
设计/竣工年份：2012年/~
建 设 地 点：浙江省宁波市东部新城金融中心南区A2-26号地块
总建筑面积：78 700 m²
总建筑高度/层数：99.8 m/23 F
结 构 形 式：框筒结构

Owner: Bank of Communication Co., Ltd.
Function: Office
Design/Completion Year: 2012/~
Construction Site: Plot A2-26, South Zone of Financial Center, Eastern New Town, Ningbo City, Zhejiang
Total Floor Area: 78,700 m²
Total Height/Floor: 99.8 m/ 23 F
Structure: Frame-Tube Structure

项目位于宁波东部新城金融中心南区A2—26号地块，用地面积为8 484 m²。项目力营造东部新城金融中心地标性建筑。建筑总图形态呈"U"字形布置。建筑形体采用方形体量穿插的手法，将玻璃的纯净感与金属杆件的垂直感理性地组合起来。这样的造型既富有流线感又具有结构美。

The project, located on plot A2-26 in South Zone of Financial Center, Eastern New Town, Ningbo, covers a site area of 8,484 m². It aims to create a landmark building in the Financial Center of Eastern New Town. The building has a U-shaped overall layout, and it is composed of stacked square volumes and realizes combination between pure glass and vertical metal rods. This results in a beautiful profile and functional structure.

山东东营市广饶县颐和国际大厦
Guangrao County Yihe International Mansion, Dongying City, Shandong

建设单位：东营渤海房地产开发有限公司
功能用途：商务、办公
设计/竣工年份：2012年/~
建设地点：山东省东营市广饶县
总建筑面积：40 180 m²
总建筑高度/层数：82.75 m/19 F
结构形式：框剪结构

Owner: Dongying Bohai Real Estate Development Co., Ltd.
Function: Commerce, Office
Design/Completion Year: 2012/~
Construction Site: Guangrao County, Dongying City, Shandong
Total Floor Area: 40,180 m²
Total Height/Floor: 82.75 m/ 19 F
Structure: Frame-Shear Wall Structure

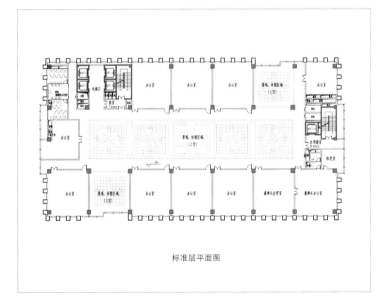

标准层平面图

一层平面图

项目位于广饶县迎宾路以北，民安路以东，正安路以西，规划用地面积为8 131 m²。项目为高端商务写字楼，裙房配备部分商业，采用智能化管理，环保节能。设计围绕着现代、简洁、高效的主题，旨在创建高品质、高人气的商业办公环境。建筑的主要功能为商业、办公。在平面布局上，追求功能区域明确、交通组织合理的布局，力图创造出人性化空间和标志性建筑。

The project is located at the north of Yingbin Road, the east of Min'an Road and the west of Zheng'an Road in Guangrao County, covering a site area of 8,131m². It is a high-end business building and some commercial functions are designed in the podium building, to realize intelligent management and environmental-friendly and energy-saving effects. The design focuses on modern, simple and efficient concept to create high-quality and welcoming office and business environment. This project mainly provides commerce and office functions. The plane layout boasts of clear functional divisions and appropriate communication organization, so as to create humanism space and landmark building.

超威电源有限公司新能源大厦
New Energy Tower of Chilwee Power Co., Ltd.

项目位于浙江省长兴市经济开发区明珠北路与发展大道交叉口。主楼内设置商业中心、内部会所、企业总部办公，部分为附属公司的租赁办公。商业中心区在主楼南侧，共4层，沿街东西向布置。会所区域设置在商业中心的上层，设有娱乐健身中心、餐饮中心、客房中心。建筑设计用矩形为单元的母题在建筑体形上演绎，在建筑主楼的西南角顶部做了不拘一格的处理，寻求未来主义的视觉感受，希望表现出建筑美学中的纯净与线条感。设计追求以比例协调、结构稳定的外形为特点的空间，用大片低反射灰白玻璃建造出水晶般明亮透彻的建筑外形。

The project is located at the crossroad of Mingzhu North Road and Fazhan Boulevard in the Economic Development Zone of Changxing City. The main building contains commercial center and internal chamber. Most of the space is used as headquarters offices and some spaces are rented to affiliated companies. The commercial center located at the southern part of the main building occupies four floors and stretches east and west along the street. Chamber is positioned above the commercial center and it contains entertainment and fitness center, dining center and guest room center. The building volume is composed of square units and the southwest top angle of the main building is treated diversely to produce a futuristic visual effect, embodying the purity and esthetic profile of the building. Reasonable proportion and outstanding appearance feature the building space, and large area of low-reflection hoar glass highlights the shining and transparent crystal appearance.

建 设 单 位：长兴超威置业有限公司
功 能 用 途：办公、商业
设计/竣工年份：2012年/~
建 设 地 点：浙江省长兴市经济开发区
总建筑面积：92 300 m²
总建筑高度/层数：100 m/26 F
结 构 形 式：框筒、框剪结构

Owner: Changxing Chaowei Real Estate Co., Ltd.
Function: Office, Commerce
Design/Completion Year: 2012/~
Construction Site: Economic Development Zone, Changxing City, Zhejiang
Total Floor Area: 92,300 m²
Total Height/Floor: 100 m/ 26 F
Structure: Frame-Tube Structure, Frame-Shear Wall Structure

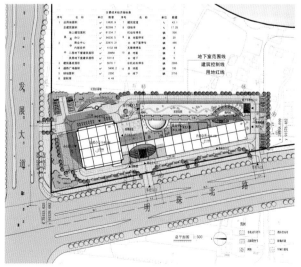

衢州国土资源局办公大楼
Quzhou State Land Resource Bureau Office Building

建 设 单 位：衢州市地产市场管理处
功 能 用 途：办公大楼
设计/竣工年份：2012年/~
建 设 地 点：浙江省衢州市西区
总 建 筑 面 积：25 000 m²
总建筑高度/层数：60 m/12 F
结 构 形 式：框架结构

Owner: Quzhou Office for Real Estate Market Management
Function: Office Building
Design/Completion Year: 2012/~
Construction Site: West District, Quzhou City, Zhejiang
Total Floor Area: 25,000 m²
Total Height/Floor: 60 m/ 12 F
Structure: Frame Structure

项目位于衢州市西区三江路以北，为了使建筑体量尽量显得挺拔，建筑主体采用典型的点式大楼形式。42m的面宽设计，在满足大楼使用舒适度要求的同时，也为北面地块预留了尽可能多的面向城市中心公园的景观视野。与此同时，对建筑细部也进行了重点考虑，在墙面划分、檐口、勒脚、构架交接处等都进行了细部处理，从而使整个建筑在简洁明快的总体风格下保持了庄重、典雅的特性。

The project is located at the north of Sanjiang Road in West District of Quzhou City. To highlight the tall volume of the building, this project adopted typical point-type tower building form. The 42 m plane width not only satisfies the comfortable space in the building, but also reserves for the north plot as large as possible landscape views towards urban central garden. Meanwhile, special attentions are paid to building details, such as division of wall, eaves, plinth and structural joint, etc., realizing a solemn, elegant and simple building image.

长兴电力调度及生产运检用房
Changxing Power Bureau Building for Power Dispatching and Production Maintenance

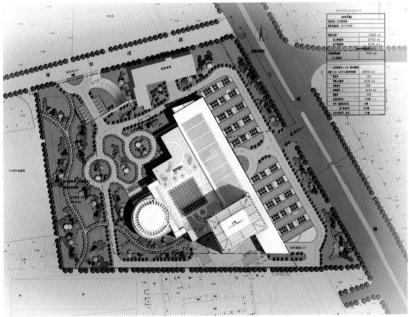

建 设 单 位：长兴电力局
功 能 用 途：电力调度、办公
设计/竣工年份：2012年/~
建 设 地 点：浙江省湖州市长兴县
总 建 筑 面 积：33 850 m²
总建筑高度/层数：66 m/15 F
结 构 形 式：框筒结构

Owner: Changxing Power Bureau
Function: Power Dispatching, Office
Design/Completion Year: 2012/~
Construction Site: Changxing County, Huzhou City, Zhejiang
Total Floor Area: 33,850 m²
Total Height/Floor: 66 m/ 15 F
Structure: Frame-Tube Structure

项目位于长兴县经济开发区，经一路西侧，占地面积20113 m²。主体包括一幢高15层的电力调度办公楼，裙房布置调度中心，辅楼布置会议中心及活动中心。设计上把主楼后退建筑控制线6 m布置，主楼东侧与控制线之间设骑楼，饰以柱廊，使室内外建筑空间自然过渡。建筑造型以整体性的界面向外界展示，使之既有面向经一路的连续街道景观，同时又能更好地塑造建筑体量，改善地块内部环境。

The project is located on the west side of Jingyi Road in the economic development zone of Changxing County, covering a site area of 20,113 m². It is composed of a 15-storey power dispatching office building, podium building used as power dispatching center and wing building providing conference center and activity center. The main building is recessed 6 m from the control line and arcade with decoration of colonnade is designed between the east side of main building and the control line, to realize transition between indoor and outdoor spaces. Integral shape of the building enjoys continuous street landscape of Jingyi Road and could create better building volume and improve internal environment of the site.

湖州市吴兴区民防建设大楼
Wuxing District Civil Defense Construction Building, Huzhou City

新建的民防指挥中心以吴兴东部新区的城市设计为基础，主要考虑建筑与环境的协调，在建筑的沿湖面采用退台的建筑形式，使得建筑形体更加活泼，与周边的山体水景等自然环境的肌理相协调。退台的设计，让建筑最大限度地迎接阳光，实现绿化。本建筑中有三个单体并存，建筑形体上相互独立，又不失整体性。立面设计延续建筑形体设计中灵活且有韵律的手法。建筑色调纯净，空间层次丰富有序，充分考虑入口、沿河等不同方向上的城市景观视觉效果。

The newly-built civil defense control center is designed on the basis of urban planning of Wuxing Eastern New District and gives full consideration to the harmony between the building and the contextual environment. The terraced appearance of the building facing to the lake infuses more vigor into the building image and realizes integration between the building and the natural landscape. The terraced design could also maximize the natural lighting and green area of the building. The building is composed of three independent volumes, which create an integral visual effect. Facade design still follows the flexible and rhythmic concept. The building uses pure color and diversified spaces to highlight the visual effect of urban landscape from different directions, such as entrance and waterfront.

建设单位：湖州吴兴南太湖建设投资有限公司
功能用途：办公
设计/竣工年份：2012年/~
建设地点：浙江省湖州市吴兴区
总建筑面积：20 482 m²
总建筑高度/层数：24 m/6 F
结构形式：框架结构
合作单位：湖州市城市规划设计研究院

Owner: Huzhou Wuxing Nantaihu Construction and Investment Co., Ltd.
Function: Office
Design/Completion Year: 2012/~
Construction Site: Wuxing District, Huzhou City, Zhejiang
Total Floor Area: 20,482 m²
Total Height/Floor: 24 m/ 6 F
Structure: Frame Structure
Cooperative Unit: Huzhou Municipal Planning and Design Institute

厦门福兴国际中心
Xiamen Fuxing International Center

建设单位：厦门市鑫福兴实业有限责任公司
功能用途：办公
设计/竣工年份：2012年/~
建设地点：福建省厦门市思明区观音山D15地块
总建筑面积：51 612 m²
总建筑高度/层数：100 m/25 F
结构形式：框剪结构

Owner: Xiamen Xinfuxing Industrial Co.,Ltd.
Function: Office
Design/Completion Year: 2012/~
Construction Site: Plot D15, Guanyin Mountain, Siming District, Xiamen City, Fujian
Total Floor Area: 51,612 m²
Total Height/Floor: 100 m/ 25 F
Structure: Frame-Shear Wall Structure

项目位于厦门市思明区03-07观音山D15地块。方案体现简约、时尚、现代、开放的特点。竖向铝型材线条的幕墙体系简洁、流畅、精致、挺拔，色调上与周围建筑物特质相吻合。底部的架空空间构成的"虚"与竖线条构成的"实"形成强烈的对比，让建筑表皮富有韵律而不失整体性。主楼体块穿插的造型使建筑具备起伏挺拔的视觉冲击力，并与城市空间相互渗透、辉映成趣。

The project is located on the plot D15 on Guanyin Mountain of Siming District 03-07, Xiamen. This scheme boasts of simple, fashionable, modern and open building image. The simple, smooth and elegant curtain wall decorated with longitudinal profiled aluminium strips responds to contextual buildings. Suspended bottom space is a "virtual" element that forms intensive contrast to the "solid" longitudinal strips, realizing rhythmic and integral building skin; stacked main building volumes produce waving and upsoaring visual impact and could be integrated into the urban space.

黄岩区政府综合服务中心
Huangyan District Governmental General Service Center

建 设 单 位：黄岩区人民政府机关事务管理局
功 能 用 途：办公
设计/竣工年份：2013年/~
建 设 地 点：浙江省台州市黄岩区北门广场
总 建 筑 面 积：59 864 m²
总建筑高度/层数：24 m/4 F
结 构 形 式：框架结构

Owner: Government Affairs Administration Bureau of Huangyan District People's Government
Function: Office
Design/Completion Year: 2013/~
Construction Site: Beimen Square, Huangyan District, Taizhou City, Zhejiang
Total Floor Area: 59,864 m²
Total Height/Floor: 24 m/ 4 F
Structure: Frame Structure

项目由四部分组成，分别为黄岩区政府广场、文体中心及会议中心、政府食堂和档案馆、广场地下停车库。造型和立面采用简洁洗练的设计手法。建筑外墙以暖灰色石材和通透玻璃为主要的表现材质。会议中心、食堂的立面与原有区政府大楼立面相互呼应，档案馆与文体中心的立面更加现代灵动。通过大块面的石材和玻璃的虚实对比，使整个建筑显得简洁大气、开放稳重。

The project is composed of four parts, including Huangyan District governmental square, cultural, sports and conference center, governmental dining hall and archives, square underground parking lot. The buildings have simple and clear shape and facade. Exterior wall is mainly decorated with warm grey stone and transparent glass. The facades of conference center and dining hall respond to that of the original regional district government building, while the facades of archives and cultural and sports center have more modern and flexible features. Strong contrast between large stone and glass panels highlights the simple, generous and open appearance of buildings.

杭州丰盛·恒泰广场
Hangzhou Fengsheng · Hengtai Plaza

项目的用地分为东西两个地块，南临郑家路，北靠潘家路，西倚榨河路，两地块以东陈路作为间隔。1号地块于郑家路和榨河路交叉处用地设置2栋21层的高层办公楼，榨河路与潘家路交叉处用地设置1栋12层的经济型酒店，再分别沿郑家路、东陈路设置5栋22层的点式办公楼。2号地块沿郑家路最东侧设置一栋20层的三星级酒店和14层办公楼的综合体；然后沿郑家路、东陈路设置4栋22层的点式办公楼。两个地块均以点板结合的形式于地块中间围合出一个中心景观庭院，作为办公空间的主中心景观。宾馆独立成区，交通和办公互不干扰。

The project is adjacent to Zhengjia Road at south, Panjia Road at north and Zhahe Road at west, and it is located on two plots, which are separated by Dongchen Road. On the No.1 plot, two 21-storey high-rise office buildings are planned at crossroad between Zhengjia Road and Zhahe Road, one 12-storey economical hotel at junction between Zhahe Road and Panjia Road, five 22-storey tower office buildings along Zhengjia Road and Dongchen Road respectively. On No.2 plot, a 20-storey 3-star hotel and a 14-storey complex office building are planned at the east end of Zhengjia Road, and four 22-storey tower office buildings along Zhengjia Road and Dongchen Road respectively. These high-rise and low-rise buildings on the two plots enclose a central landscape courtyard, which is used as the main central landscape of the offices. Hotels are concentrated in one area to realize separation between communication and office block.

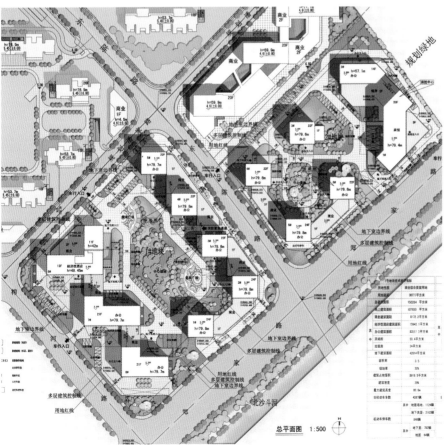

建设单位：杭州孚元泰置业有限公司
功能用途：办公、商业
设计/竣工年份：2012年/~
建设地点：浙江省杭州市三墩
总建筑面积：254 844 m²
总建筑高度/层数：80 m/22 F
结构形式：框筒结构

Owner: Hangzhou Fuyuantai Real Estate Co., Ltd.
Function: Office, Commerce
Design/Completion Year: 2012/~
Construction Site: Sandun Town, Hangzhou City, Zhejiang
Total Floor Area: 254,844 m²
Total Height/Floor: 80 m/ 22 F
Structure: Frame-Tube Structure

海正药业(杭州)有限公司总部大楼
Headquarters of Hisun Pharmaceutical (Hangzhou) Co., Ltd.

建设单位：海正药业（杭州）有限公司
功能用途：办公
设计/竣工年份：2012年/2014年
建设地点：浙江省富阳市鹿山新区
总建筑面积：50 250 m²
总建筑高度/层数：93.9 m/23 F
结构形式：框剪结构

Owner: Hisun Pharmaceutical (Hangzhou) Co., Ltd.
Function: Office
Design/Completion Year: 2012/2014
Construction Site: Lushan New District, Fuyang City, Zhejiang
Total Floor Area: 50,250 m²
Total Height/Floor: 93.9 m/ 23 F
Structure: Frame-Shear Wall Structure

项目位于富阳市鹿山新区核心区域内，接近已有城市设计中景观绿轴的临水尽端，面朝富春江，具有开阔的景观视野。本项目力求通过简洁的造型设计来体现现代企业的高效与开放，造型强调竖向肌理，以不同材质的横向划分体现层次感。裙楼在延续主楼风格的基础上，通过空间的渗透、穿插，创造出大气多变的室内环境，以满足现代化企业多变的办公需求。

This project is located in the central area of Lushan New District of Fuyang City. It is adjacent to the waterfront end of the landscape axis in the existing urban planning, facing to the Fuchun River and having wide landscape views. The design uses simple building shape to express the effective and open characteristics of the modern enterprise. Outstanding longitudinal texture and transverse division of different materials realize a sense of depth. The podium building continues the main building's features, and penetrated and intertwined spaces create diversified interior environments to satisfy different requirements of the modern enterprise.

衢州建设局办公大楼
Office Building of Quzhou Construction Bureau

建 设 单 位：衢州市政府投资项目建设中心
功 能 用 途：办公大楼
设计/竣工年份：2013年/~
建 设 地 点：浙江省衢州市西区
总 建 筑 面 积：30 650 m²
总建筑高度/层数：80 m/19 F
结 构 形 式：框架结构

Owner: Quzhou Government Investment Project Construction Center
Function: Office Building
Design/Completion Year: 2013/~
Construction Site: West District, Quzhou City, Zhejiang
Total Floor Area: 30,650 m²
Total Height/Floor: 80 m/ 19 F
Structure: Frame Structure

项目位于西区白云大道东侧，南海路北侧。设计摒弃了传统中轴对称的严肃的衙门式办公楼布局形式，建筑的整体风格为不对称的、开敞的、轻快的景观式建筑。通过石材的肌理、线条穿插对比、虚实体块的叠加等手法来体现机关大楼沉稳向上的楼宇特性。同时，设置建筑灰空间、空中花园和内庭院，使建筑的室内外空间相互融合、渗透，体现空间构成的差异性、均质性和开放性，展现出城建大楼亲民、生态的个性。

The project is located at east of Baiyun Boulevard and north of Naihai Road in West District of Quzhou City. The design discards the traditional solemn and axially symmetrical office building layout and adopts unsymmetrical, open and vigorous landscape building prototype. The stable and generous features of governmental building are successfully realized through stone texture, profile contrast and stacking up of diversified volumes. Meanwhile, grey spaces, hanging garden and interior garden are designed to blur the border between indoor and outdoor spaces, to express the difference, homogenization and openness of building space, and to highlight ecological and humanism characteristics of governmental building.

湖州电力生产运维基地
Huzhou Power Generation and Maintenance Base

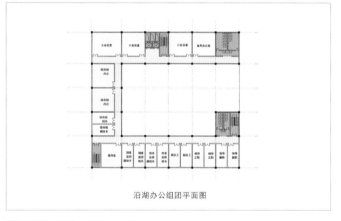

沿湖办公组团平面图

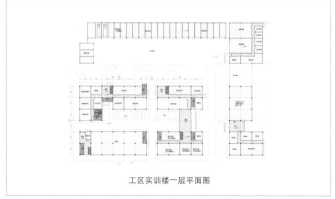

工区实训楼一层平面图

项目位于湖州市经济开发区西南分区地块，总用地面积约为100 000 m²。设计上采用"两区一轴"和"双景观核心"的总体布局。地块分为两大区块，北侧为物流工区实训区块，南侧为办公生活区块。每个区块分别设置车行入口，使得办公车流和作业特种车辆的流线得到了有效的分离。而主楼会议中心前后分别围合了入口广场和中央景观组团，即"双景观核心"。办公组团和后勤生活组团沿苕溪点状布置于南侧，占据上风向，独享场地内双景观组团和沿苕溪水景景观。

The project is located on the southwest plot in Economic Development Zone of Huzhou City, covering a total site area of about 100,000m². The design adopts the overall layout of "two zones and one axis" and "double landscape cores". The site is divided into two parts, namely the north part provides logistic and industrial functions and the south part provides office and residence. Vehicle entrance is planned for each part respectively, so as to effectively separate the flows of office workers' cars and special industrial vehicles. Entrance squares and central landscape elements are enclosed in front and at back of the conference center. Offices and logistic living facilities are dotted at south along the Shaoxi Creek. They are positioned along the upwind direction and enjoy the double landscape groups and the waterfront landscape.

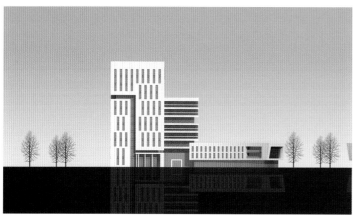

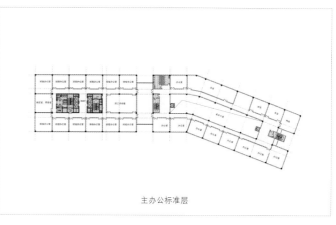

主办公标准层

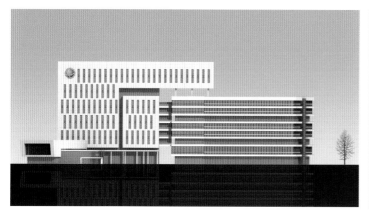

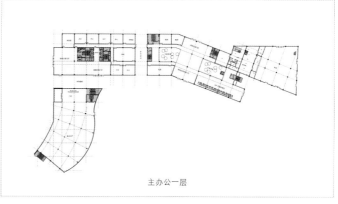

主办公一层

建 设 单 位：湖州电网	Owner: Huzhou Grid
功 能 用 途：生产运营基地	Function: Production and Operation Base
设计/竣工年份：2012年/～	Design/Completion Year: 2012/～
建 设 地 点：浙江省湖州市	Construction Site: Huzhou City, Zhejiang
总 建 筑 面 积：108 000 m²	Total Floor Area: 108,000 m²
总建筑高度/层数：50 m/12 F	Total Height/Floor: 50 m/12 F
结 构 形 式：框架结构	Structure: Frame Structure

世华大厦
Shihua Tower

建设单位：世华实业(杭州)有限公司
功能用途：办公
设计/竣工年份：2012年/2015年
建设地点：浙江省杭州市萧山区钱江世纪城
总建筑面积：84 792 m²
总建筑高度/层数：147 m/36 F
结构形式：框筒结构
合作单位：HAP萧加福建筑事务所

Owner: Shihua Real Estate (Hangzhou) Co., Ltd.
Function: Office
Design/Completion Year: 2012/2015
Construction Site: Qianjiang Century CBD, Xiaoshan District, Hangzhou City, Zhejiang
Total Floor Area: 84,792m²
Total Height/Floor: 147m/36F
Structure: Frame-Tube Structure
Cooperative Unit: HAP Architects

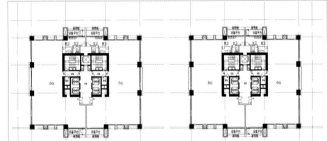

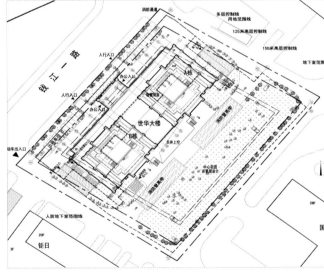

项目位于钱江南岸的钱江世纪城内，与钱江新城隔江相望。由于其地理位置特殊，区域资源丰富，其定位为杭州领先服务台商的办公大楼，为用户提供展示、接待、办公等全方位服务。该建筑以生活为基准，非短暂的风尚理念寓意于新古典主义的设计手法中，细致的雕花图案、跃动的拱窗、繁复的线条无不体现其高贵的气质。古典建筑语汇加上现代空间的布局，典雅氛围浑然天成。

The project is located in the Qianjiang Century CBD by the south bank of Qianjiang River and faces to the Qianjiang New Town across the river. Enjoying special geographical location and abundant regional resources, it is planned as the first office building for Taiwan companies, containing various functions, such as exhibition, reception, offices, etc. This life-based building uses neoclassic design method to express sustainable fashion and to show its elegant appearance in delicate carved patterns, arched windows and complex profile. Classical architectural language is perfectly integrated with modern spatial layout, realizing a natural and elegant office building.

慈溪龙山镇灵峰路综合写字楼
General Office Building on Lingfeng Road of Longshan Town, Cixi

建 设 单 位：慈溪市慈东城镇建设投资开发有限公司
功 能 用 途：办公
设计/竣工年份：2012年/~
建 设 地 点：浙江省慈溪市龙山新城
总 建 筑 面 积：43 324 m²
总建筑高度/层数：36.3 m/9 F
结 构 形 式：框架结构

Owner: Cixi Cidong Urban Construction Investment and Development Co., Ltd.
Function: Office
Design/Completion Year: 2012/~
Construction Site: Longshan New Town, Cixi City, Zhejiang
Total Floor Area: 43,324 m²
Total Height/Floor: 36.3 m/ 9 F
Structure: Frame Structure

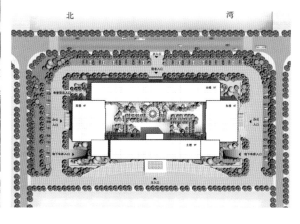

项目位于慈溪市龙山新城行政中心核心区的中心地块，面朝灵峰浦，东南面为规划中的游憩集会广场，西北面为规划的清湾北路，背山面水，环境优美。设计采用"四水归堂"的理念，意在用现代的建筑语言演绎现代写字楼的办公方式，创造多方位、多层次的交流空间。建筑延续江南典型的平面布局方式，由四个单体相连围合而成，构成传统的四合院布局。在立面造型设计上以"堂正"为出发点，即端正堂皇。竖向线条的挺拔，玻璃幕墙的明亮通透，使整个建筑体亦刚亦柔，堂正婉约。

The project is located on a central plot in the administrative center of Longshan New Town, Cixi. It faces to the Lingfeng River and is adjacent to a public square which is in planning at southeast, Qingwan North Road which is under planning at northwest, enjoying beautiful mountainous and waterfront landscape. The design is based on the concept of "influx of four rivers" and uses modern architectural languages to interpret the features of modern office building and to create diversified communication spaces. The four independent buildings adopt the representative plan layout which is popular in South of Yangtze River and enclose a traditional courtyard layout. The facades are decorated with solemn and elegant features and boast of masculine longitudinal lines and transparent glass curtain wall, creating an integrated building image of tenderness and elegance.

诸暨城东中心区A2地块综合办公大楼
General Office Building on Plot A2 of Eastern Central District, Zhuji

　　A2地块设计包括三幢"品"字形布局的高层及其北侧弧形裙房。三幢高楼均为综合办公，裙房部分安排会议中心及后勤餐饮等功能。在东、南、西、北四个方向均设置了两个出入口。在建筑造型上，结合地块内建筑的横向"品"字形布局及北侧裙房的弧形展开，以充满韵律感且富于变化的刚性竖线条分隔来塑造建筑的挺拔感，以朴素儒雅的建筑色彩为基调，强调建筑的纯净庄重。

Plot A2 is planned to build three high-rise buildings which are distributed in a triangular layout and an arched podium building at the north. These high-rise buildings will be office spaces and the podium building will provide various functions, such as conference center, logistics, catering, and so on. Two entrances are set at the east, south, west and north respectively. Based on transverse distribution of buildings and large arched plan area of podium building, rhythmical and diversified rigid vertical division strips are designed to create generous building appearance. The purity and solemnity of the buildings are highlighted through simple and elegant colors.

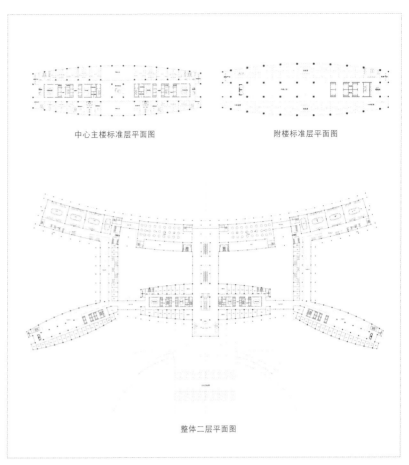

中心主楼标准层平面图　　附楼标准层平面图

整体二层平面图

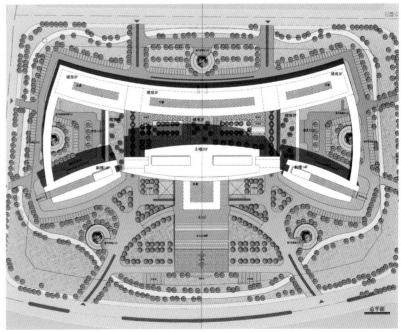

建设单位：诸暨市城市建设投资发展有限公司　　Owner: Zhuji Urban Investment and Development Co., Ltd.
功能用途：办公　　Function: Office
设计/竣工年份：2012年/~　　Design/Completion Year: 2012/~
建设地点：浙江省诸暨市　　Construction Site: Zhuji City, Zhejiang
总建筑面积：200 000 m²　　Total Floor Area: 200,000 m²
总建筑高度/层数：98 m/20 F　　Total Height/Floor: 98 m/20 F
结构形式：框剪结构　　Structure: Frame-Shear Wall Structure

正凯中心
Zhink Center

项目为超高层办公综合体，由一栋150 m高的A楼和一栋46.45 m高的B楼组成双子塔楼。A楼坐北朝南，采用偏心筒布局，在中央形成完整的大空间，并保证办公空间全部南向布置。B楼相对正南北向偏转，既避开A楼遮挡，也向内环抱着绿色中庭，将多层次的景观环境带入建筑空间。建筑造型上利用微微收分的曲线避免了体量的臃肿感，建筑形体高低错落，造型简洁明快，主立面比例匀称，侧立面高耸挺拔，整个建筑形成强烈的上升气势。

This project is a superhigh business complex which is composed of 150 m high building A and 46.45 m high building B. Building A faces to the south and adopts an eccentric tube layout to form a large central space and to guarantee that all the offices face to the south. Building B has a deflection angle from the normal south and north to avoid the shade projected by Building A and to enclose a green atrium, introducing multi-tiered landscape environment into building. The building adopts a slightly contracted curve image to avoid clumsy feeling. The building volume has different elevations and simple and clear appearance. The main facade has proper proportion and the side facade is tall and straight, creating a rising image of the whole building.

建设单位：浙江正凯置业有限公司
功能用途：办公
设计/竣工年份：2012年/~
建设地点：浙江省杭州市萧山区
总建筑面积：106 826 m²
总建筑高度/层数：150 m/37 F
结构形式：框筒结构
合作单位：深圳市平行设计顾问有限公司

Owner: Zhejiang Zhink Real Estate Co., Ltd.
Function: Office
Design/Completion Year: 2012/~
Construction Site: Xiaoshan District, Hangzhou City, Zhejiang
Total Floor Area: 106,826 m²
Total Height/Floor: 150 m/ 37 F
Structure: Frame-Tube Structure
Cooperation Unit: Shenzhen Parallel Design Consultants Co., Ltd.

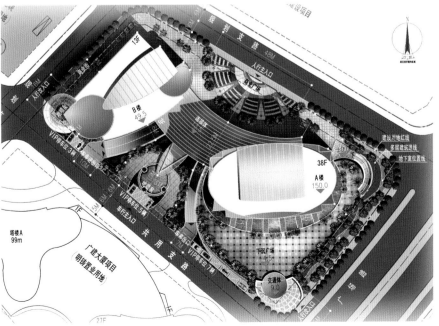

三宏国际大厦
Sanhong International Building

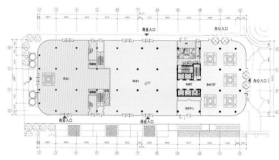

一层平面图

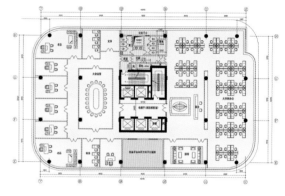

办公标准层

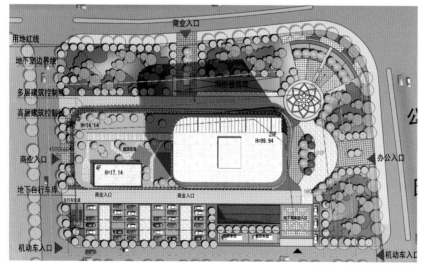

项目地处钱江世纪城公园西路和广场路交会处，旁边即为地铁站，无论是地理位置还是景观条件都非常优越。整个基地南北长约67 m，东西宽约125 m，地块呈东西走向，总用地面积约8 503 m²，是包括高级办公楼、商业楼以及两层地下室在内的综合建筑群。项目占地面积2 551.5 m²，地上建筑面积为34 012 m²，地下建筑面积14 207 m²。

This project is located at the junction between the Park West Road of Qianjiang Century CBD and Square Road. The metro station is nearby. It has both favorable geographical and landscape conditions. The length from the north to the south of the base is 67 m, and its width is 125 m. The area goes from the east to the west. The total area is 8,503 m². The comprehensive buildings include advanced office buildings, commercial buildings, and two-floor basement. The total covering area is 2,551.5 m²; the construction area above the ground is 34,012 m²; and the construction area under the ground is 14,207 m².

西立面

南立面

东立面

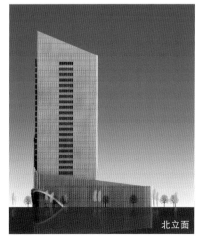

北立面

建设单位：杭州三宏房地产有限公司	Owner: Hangzhou Sanhong Real Estate Co., Ltd.
功能用途：商业、办公	Function: Commerce, Office
设计/竣工年份：2011年/2014年	Design/Completion Year: 2011/2014
建设地点：浙江省杭州市钱江世纪城	Construction Site: Qianjiang Century CBD, Hangzhou City, Zhejiang
总建筑面积：48 219 m²	Total Floor Area: 48,219 m²
总建筑高度/层数：120 m/25 F	Total Height/Floor: 120 m/ 25 F
结构形式：框筒结构	Structure: Frame-Tube Structure

杭州和茂大厦
Hangzhou Hemao Building

建设单位：杭州经济技术开发区资产经营集团有限公司
功能用途：办公
设计/竣工年份：2011年/2013年
建设地点：浙江省杭州市经济技术开发区幸福北路
总建筑面积：64 884.4 m²
总建筑高度/层数：97.46 m/25 F
结构形式：框剪结构

Owner: Hangzhou Economic and Technological Development Zone Assets Operation Group Co., Ltd.
Function: Office
Design/Completion Year: 2011/2013
Construction Site: Xingfu North Road in Economic and Technological Development Zone, Hangzhou City, Zhejiang
Total Floor Area: 64,884.4 m²
Total Height/Floor: 97.46 m/25 F
Structure: Frame-Shear Wall Structure

项目位于杭州市下沙经济技术开发区内，基地呈L形，西半部面积较大，并呈近似正方形。主楼坐北朝南，位于西半部的西侧，中偏南位置，附楼紧贴主楼东侧成T形布置。整个建筑由两个高低错落的体块穿插而成，建筑形体节节升高，形成丰富的城市空间和城市天际线，主体建筑直接落地，建筑立面显得更加挺拔、简洁与庄重。

The project is located in Xiasha Economic and Technological Development Zone, Hangzhou City. The L-shaped site has a larger west part which is approximately square. The main building faces to the south and stands on the west side and somewhat south of the west part; attached building leans on the east side of the main building and is planned in a T-shaped layout. The whole building is composed of two volumes of different heights, creating diversified urban spaces and skylines. Main body of the building stands on the ground, boasting of tall, simple and solemn facades.

麦道大厦
Maidao Mansion

建设单位：杭州麦道置业有限公司
功能用途：办公、商业
设计/竣工年份：2011年/~
建设地点：浙江省杭州市临平
总建筑面积：40 800 m²
总建筑高度/层数：60 m/13 F
结构形式：框架结构

Owner: Hangzhou Maidao Real Estate Co., Ltd.
Function: Office, Commerce
Design/Completion Year: 2011/~
Construction Site: Linping, Hangzhou City, Zhejiang
Total Floor Area: 40,800 m²
Total Height/Floor: 60 m/ 13 F
Structure: Frame Structure

项目位于杭州副城临平新城南苑街道，地块周边生活、旅游配套集中，商业氛围浓厚，交通极为便利。项目总建筑面积40 800 m²，占地面积约3 085 m²，主要用途为办公、商业、餐饮及服务建筑，地上13层，地下3层，主体高60 m。

The project is located on the Nanyuan Street of Linping New City, a sub-city of Hangzhou. The site enjoys concentrated living and tourism resources, intensive commercial atmosphere and convenient communication. It has a gross building area of 40,800 m² and covers a site area of about 3,085 m². The main functions include offices, retail spaces, restaurant and service facility, etc. The mansion is composed of 13 overground floors and 3 basement floors, and the height of its main body is 60 m.

台州中央商务区3—02,04地块设计
Taizhou CBD Plots 3-02 and 04

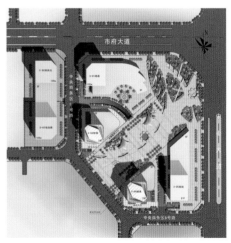

建设单位：浙江省台州经济开发区经济发展总公司
功能用途：办公
设计/竣工年份：2011年/~
建设地点：浙江省台州市中央商务区
总建筑面积：80 000 m²
总建筑高度/层数：98 m/24 F
结构形式：框筒结构

Owner: Zhejiang Taizhou Economic Development Zone Economic Development General Company
Function: Office
Design/Completion Year: 2011/~
Construction Site: Central Business District, Taizhou City, Zhejiang
Total Floor Area: 80,000 m²
Total Height/Floor: 98 m/24 F
Structure: Frame-Tube Structure

项目位于台州市中心区主轴线几何中心，中央商务区的东北角，是集商务办公、金融保险、酒店商业、休闲公园于一体的城市活动中心。其中03—02地块为台州开发区经济发展总公司办公大楼，03—04地块为浙江泰隆商业银行大楼。主体塔楼朝中轴线作45°方向切角处理，与裙房自然衔接，局部竖向玻璃幕墙结合顶部"钻石"形象进行处理，与城市主轴空间产生对话，形成"双子座"门户建筑意象。

The project is located at the northeast corner in the Central Business District, which is the geometric center along the main axis of Taizhou, and it is an urban center integrating diversified functions, such as business office, financial insurance, hotel and leisure park, etc. In this project, an office building of Taizhou Economic Development Zone Economic Development General Company is planned on plot 03-02 and Zhejiang Tailong Commercial Bank building is planned on plot 03-04. The main tower is chamfered with a 45° angle to the axial line, and it is naturally connected with the podium building. Local longitudinal glass curtain wall is connected with the (diamond) roof to have a dialogue with urban space along the main axis and to produce a portal image of twin towers.

阿里巴巴B2B二号园区
Alibaba B2B No.2 Park

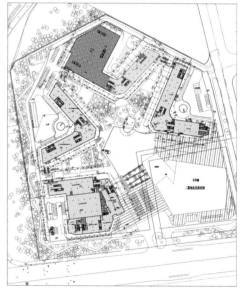

建筑的外形衍生自其独特的平面布局。为适合基地不规则的五边形地貌，同时围合出尽量大的共享中央广场，建筑的平面呈现蜿蜒的布局，沿基地周边展开。建筑的整体气势决定了建筑外立面的处理手法，即建筑的外皮必须配合平面和大体量的总体趋势，必须加强并补充建筑向水平方向延展的不可抗拒的力量，使相互之间用骑楼相连接的4栋单体建筑能够真正形成一个气势磅礴的整体。建筑外观简洁大方，舒展轻盈。建筑立面强调横向线条，通过均匀的外墙材质刻意地模糊楼层之间的分隔，使建筑以饱满的体量和整体的气势突出于城市肌理之中。

The building appearance is designed according to its unique plan layout. The building has a sinuous plan layout unfolding along the site borders, with the purposes to take advantage of irregular pentagonal terrain and to enclose a shared central square as large as possible. Overall image of the building determines the treatment method of the building facade, namely building skin must follow the general trend of plane and large volume, and the resistance against horizontal expansion must be strengthened. Four independent buildings are connected by arcade to form a true integral volume. Building adopts simple and generous appearance to create a stretching and lightweight image. Building facades are mainly decorated with transverse lines. Division among floors is properly blurred through uniform exterior wall materials, creating an integral building image in the urban context.

建设单位：阿里巴巴（中国）网络技术有限公司
功能用途：办公
设计/竣工年份：2011年/~
建设地点：浙江省杭州市滨江区
总建筑面积：195 700 m²
总建筑高度/层数：34 m/7 F
结构形式：框架结构
合作单位：香港王董国际有限公司

Owner: Alibaba.com (China) Technology Co., Ltd.
Function: Office
Design/Completion Year: 2011/~
Construction Site: Binjiang District, Hangzhou City, Zhejiang
Total Floor Area: 195,700 m²
Total Height/Floor: 34 m/ 7 F
Structure: Frame Structure
Cooperation Unit: Wong & Tung International Limited

杭州汇鑫大厦
Hangzhou Huixin Mansion

建设单位：杭州环东置业有限公司
功能用途：办公
设计/竣工年份：2011年/~
建设地点：浙江省杭州市钱江世纪城
总建筑面积：89 100 m²
总建筑高度/层数：98.5 m/25 F
结构形式：框剪结构
合作单位：5+1五合国际

Owner: Hangzhou Huandong Real Estate Co., Ltd.
Function: Office
Design/Completion Year: 2011/~
Construction Site: Qianjiang Century CBD, Hangzhou City, Zhejiang
Total Floor Area: 89,100 m²
Total Height/Floor: 98.5 m/25 F
Structure: Frame-Shear Wall Structure
Cooperation Unit: 5+1 Werkhart International

项目地处杭州市钱江世纪城，周边未来规划的项目规模较大，高度较高，结合对比分析本项目较难形成突出的体量，同时周边已开工项目多为双塔或双板，为了突出自身的特点，建筑形象应寻求差异化。对规划条件及周边环境因素进行综合考虑，主楼造型以下小上大的三叶形玻璃体块为特征，突出璀璨外形，彰显个性形象，同时因体形的变化以获得最大的观景面，使其三面都能观江。

The project is located in Qianjiang Century CBD of Hangzhou City and it is surrounded by large-scale and high projects which will be planned in the future. Based on contrastive analysis, it is relatively difficult to create an outstanding building volume in this project; meanwhile, most contextual projects under construction are twin towers or twin-slab buildings, so the building shall have a differential appearance to stand out from the context. After general consideration of planning conditions and surrounding environmental factors, the main building adopts a trefoil glass volume to highlight the distinguished appearance and image and to realize maximum landscape view towards the river on three facades.

国金中心
International Financial Center

一层平面图

避难层平面图

建设单位：杭州港基房地产开发有限公司
功能用途：写字楼、办公
设计/竣工年份：2010年/~
建设地点：浙江省杭州市钱江世纪城
总建筑面积：76 366 m²
总建筑高度/层数：129.8 m/28 F
结构形式：框剪结构

Owner: Hangzhou Gangji Real Estate Development Co., Ltd.
Function: Business Building, Office
Design/Completion Year: 2010/~
Construction Site: Qianjiang Century CBD, Hangzhou City, Zhejiang
Total Floor Area: 76,366 m²
Total Height/Floor: 129.8 m/ 28 F
Structure: Frame-Shear Wall Structure

项目位于杭州市钱江二路北侧钱江世纪城I-2-03地块，为超高层写字楼。项目以独栋板楼形式布局，建筑主楼迎面朝向东南方向，主要入口广场与钱江二路紧密结合布置，主体建筑退让26.8 m，底层局部凸出，形成1 500 m²左右的入口广场，同时也是主要的集散广场。在地块西北以及东北部设局部的采光庭院，既丰富了场地布置，在功能上又为地下室提供采光。

The project is located on plot I-2-03 of Qianjiang Century CBD at north of Qianjiang No.2 Road in Hangzhou City and it is super-high business building. This project adopts singe-family slab building layout. The main facade faces to the southeast and main entrance square is connected with Qianjiang No.2 Road. The building body is recessed 26.8 m and local position on the ground floor protrudes to form an entrance square of about 1,500 m², which is used as a main public square. Local lighting courtyards are planned on northwest and northeast parts of the site, not only enriching the site layout, but also providing natural light to the basement.

杭州华联钱塘会馆
Hangzhou UDC Qiantang Chamber

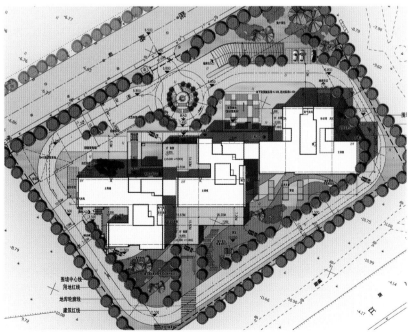

建设单位：杭州华联置业有限公司
功能用途：办公
设计/竣工年份：2010年/2013年
建设地点：浙江省杭州市钱江新城剧院路
总建筑面积：62 110 m²
总建筑高度/层数：81.8 m/21 F
结构形式：框剪结构
合作单位：澳大利亚柏涛（墨尔本）建筑设计有限公司亚洲分公司

Owner: Hangzhou UDC Real Estate Co., Ltd.
Function: Office
Design/Completion Year: 2010/2013
Construction Site: Juyuan Road, Qianjiang New Town, Hangzhou City, Zhejiang
Total Floor Area: 62,110 m²
Total Height/Floor: 81.8 m / 21 F
Structure: Frame-Shear Wall Structure
Cooperation Unit: Peddle Thorp Architects (Melbourne) Asian Branch

项目地处钱江新城中心商务区核心区，平面上一共设计了三个单元空间，错列排布，彼此之间以公共平台相连接，共同组成一栋完整的建筑物。立面的设计着重表达了滨江建筑的意象，舒展的横向线条简洁大方，富有现代气息。以双层玻璃幕墙追求简约与高效，局部的体块变化丰富了沿江视觉效果，并呈现出硬朗、明快的矩形形态，以谦逊的姿态沿江分布。该项目成为该地段一个崭新的人文景观，为钱塘江的景色再添巧妙的一笔。

The project, located in the central area of CBD district of Qianjiang New Town, is composed of three stagger volumes, which are connected by public platform to form an integral building. The facade design highlights the characteristics of waterfront building, namely simple and generous extended transverse profile embodies modern architectural trend. Double-layer glass curtain wall focuses on simplicity and high efficiency and diversified local volumes enrich the waterfront visual effects and create a vivid rectangular image. The building faces to the water and its modest image becomes a brand-new humanism landscape on the site, infusing great vigour into the waterfront landscape.

温岭九龙商务中心（办证中心）
Wenling Jiulong Business Center (Certificate Service Center)

建 设 单 位：温岭九龙汇开发建设有限公司
功 能 用 途：商业、办公
设计/竣工年份：2010年/~
建 设 地 点：浙江省温岭市
总 建 筑 面 积：31 700 m²
总建筑高度/层数：27.9 m/6 F
结 构 形 式：框架结构

Owner: Wenling Jiulonghui Real Estate Development Co., Ltd.
Function: Commerce, Office
Design/Completion Year: 2010/~
Construction Site: Wenling City, Zhejiang
Total Floor Area: 31,700 m²
Total Height/Floor: 27.9 m/ 6 F
Structure: Frame Structure

项目的功能从竖向上分成三个部分：一至三层为办证大厅及相关配套用房，第四层为招投标区，五至六层为综合办公区。办证大厅的交通和招投标区的垂直交通独立设置。中心的外观造型，以底层斜向的三层高的入口空间为核心展开，采用石材和窄窗的外墙立面体现稳重、大方的建筑基调。建筑整体雕塑感强烈，四个立面各具"表情"，展现了一个具有地标性、稳重开放、具有亲和力的新型办公建筑形象。

This project is required to distribute its functions into three parts vertically, namely 1F-3F are certificate service lobby and related supporting facilities, 4F is bidding service space, 5F and 6F are general office block. Meanwhile, vertical communication of the certificate service lobby and bidding service space are designed independently. The central oblique 3-storey entrance on ground floor is the core element of the building; stone facade decorated with narrow windows highlights the elegant and generous appearance of the building. The sculpture volume has different "faces" on four facades, embodying the landmark image of this new office building.

青川县行政中心
Qingchuan Administration Center

建 设 单 位：青川县城乡规划建设和住房保障局
功 能 用 途：行政办公
设计/竣工年份：2010年/2013年
建 设 地 点：四川省青川县
总 建 筑 面 积：53 383 m²
总建筑高度/层数：29.4 m/7 F
结 构 形 式：框架结构

Owner: Qingchuan Bureau of Rural Planning Construction and Housing Security
Function: Administrative Office
Design/Completion Year: 2010/2013
Construction Site: Qingchuan County, Sichuan
Total Floor Area: 53,383 m²
Total Height/Floor: 29.4 m/ 7 F
Structure: Frame Structure

项目是"5.12"汶川大地震后的灾后重建项目，基地位于青川县城核心区域。整个项目包括行政主楼、政务中心、公共服务中心、财经中心、农业中心、档案馆六个子项。项目整体布局强调空间的严谨、有序、开放，两条十字交叉的轴线成为整个建筑群的骨架，串连起地块内的各个建筑。建筑体形方正，立面风格上强调虚实对比以及立面肌理的韵律感。项目整体风格体现了政府办公建筑应有的庄重、大方。

The project is a reconstruction project after "5.12" Wenchuan Earthquake and it is located in the central district of Qingchuan County. It consists of six functional buildings, including main administrative building, governmental affair center, public service center, financial and economic center, agricultural center and archives, etc. The overall layout focuses on regular and open spaces and two crossed axes acting as skeleton of the whole building cluster connects all buildings together. The buildings have square shape and facades boast of contrast between virtual and solid elements, as well as rhythmical appearance. The overall building features embody the solemnity and generosity of governmental office building.

宁波银行总部大厦
Bank of Ningbo Headquarters

建 设 单 位：宁波银行股份有限公司
功 能 用 途：办公
设计/竣工年份：2010年/～
建 设 地 点：浙江省宁波东部新城金融南区A2-25号地块
总 建 筑 面 积：104 907 m²
总建筑高度/层数：141.31 m/31 F
结 构 形 式：框筒结构
合 作 单 位：德国GMP建筑设计事务所

Owner: Bank of Ningbo Co., Ltd.
Function: Office
Design/Completion Year: 2010/~
Construction Site: Plot A2-25 in Financial South District, East New Town, Ningbo City, Zhejiang
Total Floor Area: 104,907 m²
Total Height/Floor: 141.31 m/ 31 F
Structure: Frame-Tube Structure
Cooperation Unit: GMP Architects

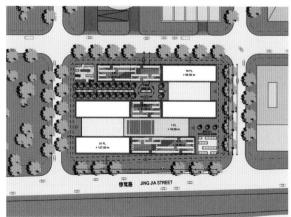

本项目在2010年国际投标中获第一名。建筑地下三层，地上两座塔楼，其中一座为超高层。项目通过一系列的建筑手法和语汇营造出了一个具有张力的建筑群体，通过两座塔楼和三段板式裙楼在建筑高度上的错落处理，使建筑本身的造型丰富多样。因为特殊的造型手法而形成的建筑体块之间的"缝"一来可以为银行营业厅提供场所，二来也为室外的景观处理带来新的亮点。一条景观带将西面的城市花园引入地块内部，同时将建筑体量分割得更加合理。

This is the winning proposal in a 2010 international bid competition. The building is composed of 3-storey basement and two tower buildings, one of which belongs to superhigh building. A series of architectural languages and design methods are adopted to produce an outstanding building cluster. Different elevations of the two tower buildings and 3-segment slab podium realize diversified building shapes. The "gap" generated among building volumes due to special design methods could not only provide bank public space, but also infuse new element into the outdoor landscape belt. A piece of landscape belt introduces the west urban garden into the site and realizes more reasonable division of building volume.

钱江新城金融中心
Qianjiang New Town Financial Center

建设单位：中国工商银行股份有限公司浙江省分行营业部
　　　　　华融金融租赁股份有限公司
　　　　　浙江新华期货经纪有限公司
　　　　　浙商银行股份有限公司
功能用途：办公
设计/竣工年份：2011年/～
建设地点：浙江省杭州市钱江新城
总建筑面积：291 822 m²
总建筑高度/层数：150 m/34 F
结构形式：框筒+钢结构
合作单位：德国GMP建筑设计事务所

Owner: ICBC Zhejiang Branch/ China Huarong Financial Leasing Co., Ltd. /Zhejiang Xinhua Futures Brokerage Co., Ltd./ China Zheshang Bank.
Function: Office
Design/Completion Year: 2011/～
Construction Site: Qianjiang New Town, Hangzhou City, Zhejiang
Total Floor Area: 291,822 m²
Total Height/Floor: 150 m/ 34 F
Structure: Frame-Tube and Steel Structure
Cooperative Unit: GMP Architects

项目位于杭州市钱江新城核心区。三个地块由四家建设单位共同开发，目的在于打造一个整体和谐而又彰显个性的高端金融区块。设计的目标是雕琢一个有共性的楼群，就是要在整体性和多样性之间找到平衡。我们将众多相同的元素作了有趣的不同组合，用简单的手法塑造了各建筑不同的个性。在整个组合中，使每个成员更加强调自我。竖向耸立的石材外墙赋予了楼群稳重但雅致的整体形象，给人更多的信任感，并且提供了最大化的观景视线。

This project is located in the central district of Qianjiang New Town in Hangzhou City. Three plots are jointly developed by four construction units with the purpose to create an integral and harmonious high-end financial zone. The design aims to create a unified building cluster, in which to realize balance between integrity and diversity. A large number of same elements are combined in different and interesting ways, so as to infuse different characteristics into each building by using a simple means. This vision could enable users to find their own unique features. Longitudinal stone exterior wall embodies the elegant overall image of the building cluster, providing more confidence and maximum landscape view for users.

杭州日信国际中心
Hangzhou Rixin International Center

建 设 单 位：日信置业有限公司
功 能 用 途：办公
设计/竣工年份：2010年/~
建 设 地 点：浙江省杭州市钱江新城
总 建 筑 面 积：62 566 m²
总建筑高度/层数：100 m/21 F
结 构 形 式：框剪结构

Owner: Rixin Real Estate Co., Ltd.
Function: Office
Design/Completion Year: 2010/~
Construction Site: Qianjiang New Town, Hangzhou City, Zhejiang
Total Floor Area: 62,566 m²
Total Height/Floor: 100 m/ 21 F
Structure: Frame-Shear Wall Structure

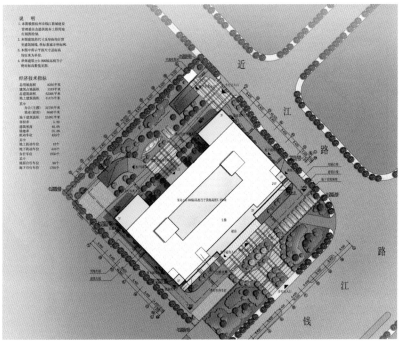

　　项目位于钱江新城一期规划范围内，是一座集大型商场、超市、金融、办公、休闲娱乐和停车配套等多项功能于一体的综合性大楼。庭院的围廊式平面有别于传统的单廊、复廊或者外廊式平面，不仅兼具以上几种形式的优点，做到经济、便利、私密和生态，而且打破了建筑室内外界线，丰富了建筑内部的空间构成，大大提高了空间品质。为加强内庭院和外环境横向的空气流通，于每层的东面设了一个开口，通过直接向外部空间开敞的空中花园实现渗透。

The project, located in the planning scope of Qianjiang New Town Phase I, is a complex building providing diversified functions, such as large shopping mall, supermarket, finance, office, leisure, entertainment and parking, etc. Peristyle plane layout is different from traditional single-corridor, double-corridor or external-corridor plane layouts, because it is not only economical, convenient, private and ecological, which are also realized in the aforesaid layout forms, but also blurs the border between indoor and outdoor spaces, diversifies interior spaces and greatly improves space quality. With the purpose to strengthen transverse air ventilation between internal courtyard and external environment, one opening is designed on east facade on each floor to realize connection with the open hanging garden.

潮峰钢构集团有限公司幕墙生产车间改造工程
Curtain Wall Production Plant Reconstruction of Triumpher Steel Structure Group Co., Ltd.

建设单位：潮峰钢构集团有限公司
功能用途：办公
设计/竣工年份：2013年/2014年
建设地点：浙江省杭州市萧山经济技术开发区
总建筑面积：40 126 m²
总建筑高度/层数：16.9 m/3 F
结构形式：钢结构

Owner: Triumpher Steel Structure Group Co., Ltd.
Function: Office
Design/Completion Year: 2013/2014
Construction Site: Economic and Technological Development Zone of Xiaoshan District, Hangzhou City, Zhejiang
Total Floor Area: 40,126 m²
Total Height/Floor: 16.9 m/3 F
Structure: Steel Structure

　　项目位于杭州市萧山区原潮峰钢构集团厂区，改造内容为幕墙生产车间平面功能改造及整个厂区立面改造。
　　平面改造内容：原三层幕墙生产车间建筑呈矩形布置，一层、二层为幕墙加工车间，三层为集团办公室。现设计保持原建筑体量不变，一至三层功能改为办公室，并补齐原建筑东北角的缺口。由于原厂房单层面积较大，作为办公楼在采光及交通组织上存在较多不利之处，因此考虑在原建筑内部设置一层至三层连通的中庭，以中庭为中心组织内部交通，并且能有效改善内部办公的采光。潮峰钢构集团厂区改造后将建成全新的萧山空间结构科技产业园，成为集办公、商业、餐饮于一体的综合体。

The project is located in the factory of former Triumpher Steel Structure Group in Xiaoshan District, Hangzhou City, and it includes rebuilding plane function of the curtain wall production plant and all facades.
Plane layout reconstruction: the original 3-storey curtain wall production plant is distributed in a rectangular layout. 1F and 2F are used for curtain wall production and 3F is used as offices of Triumpher. The original building volume will not be changed, but 1F-3F are rebuilt as offices and the northeast gap of the original building will be repaired. Considering the large area of single floor and multiple inconveniences of office building in lighting and communication, a 3-storey through atrium is designed in the original building and internal communication is planned around the atrium, to obviously improve the interior lighting effect. After reconstruction, the Triumpher Steel Structure Group factory will become a brand-new space structure scientific and technological industry park in Xiaoshan District, and a complex building integrating office, retail and restaurant functions will be created.

江西德兴市德兴大厦
Jiangxi Dexing Mansion

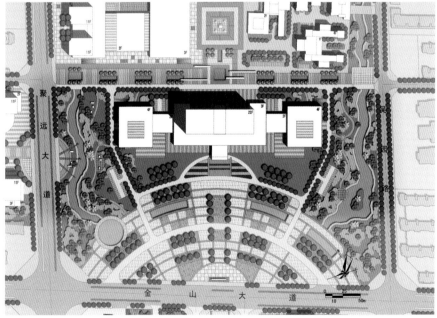

建 设 单 位：德兴市建设局
功 能 用 途：商业、办公
设计/竣工年份：2010年/2014年
建 设 地 点：江西省德兴市
总 建 筑 面 积：100 000 m²
总建筑高度/层数：81.5 m/19 F
结 构 形 式：框剪结构

Owner: Dexing Construction Bureau
Function: Commerce, Office
Design/Completion Year: 2010/2014
Construction Site: Dexing City, Jiangxi
Total Floor Area: 100,000 m²
Total Height/Floor: 81.5 m/19 F
Structure: Frame-Shear Wall Structure

项目地处德兴市中央商务区核心位置。方案尊重原有地形、地貌的特征，将规划布局、建筑设计与周围城市环境设计有机地结合起来。主体建筑在功能上包括相对独立的办公楼和附楼以及为两者服务的会议室部分。在立面处理上，采用垂直向上的竖向线条作为这栋建筑形态构思的原型和基调，从而赋予它积极、稳重、大气的形式。主体建筑的体形流畅而富于动势，石材的运用突出建筑稳重的个性，南北两侧的幕墙轻盈地向上升起，似展开的双翼，寓意为城市历史性的腾飞。

The project is located at the central position in CBD of Dexing City. The design scheme respects existing terrain and geological conditions, and organically combines planning layout, architectural design and urban context together. The main body includes diversified functions, such as relatively independent offices, service auxiliary building and conference space. The facades use vertical and upsoaring architectural language to express its design prototype and concept of this building and to infuse positive, solemn and generous features into the building. The smooth and dynamic main building body is decorated with stone materials to highlight its elegant image, curtain wall systems on both south and north facades are just like wings embodying historical boom of the city.

余姚商会大厦
Yuyao Chamber of Commerce Building

建设单位：余姚市总商会
功能用途：办公
设计/竣工年份：2009年/2012年
建设地点：浙江省余姚市
总建筑面积：59 473 m²
总建筑高度/层数：88 m/23 F
结构形式：框筒结构

Owner: Yuyao General Chamber of Commerce
Function: Office
Design/Completion Year: 2009/2012
Construction Site: Yuyao City, Zhejiang
Total Floor Area: 59,473 m²
Total Height/Floor: 88 m/ 23 F
Structure: Frame-Tube Structure

项目位于余姚市市中心，由两栋23层高的双子楼——南楼和北楼组成。主楼之间的连接裙房为由南向北的连续界面，与城市肌理融合。基地的最北侧布置为地面停车场，位于公共绿地之前，最大限度地减少了对环境的干扰。南楼和北楼分别位于控制红线的最北端和最南端，中间形成了低矮的四层裙房，有利于形成东侧南雷高架桥与西侧中山河的景观联系，满足规划设计的要求。

The project, located in the urban center of Yuyao City, is composed of 23-storey twin towers standing at north and south on the site respectively. The podium building connecting the main buildings forms a continuous interface stretching from south to north and responding to the urban context. A ground parking lot is planned at the northmost part on the site and faces to a public green land, greatly reducing impact on the environment. The two buildings stand respectively at the northmost and southmost ends of the control lines and enclose a 4-storey podium building. Such layout could realize connection between Nanlei Viaduct at east and Zhongshan River at west, so as to satisfy the planning and design requirements.

杭州协和大厦
Hangzhou Xiehe Building

建设单位：杭州协和辉丰房地产开发有限公司
功能用途：办公
设计/竣工年份：2010年/~
建设地点：浙江省杭州市萧山区
总建筑面积：95 030 m²
总建筑高度/层数：160 m/34 F
结构形式：框筒结构

Owner: Hangzhou Xiehe Huifeng Real Estate Development Co., Ltd.
Function: Office
Design/Completion Year: 2010/~
Construction Site: Xiaoshan District, Hangzhou City, Zhejiang
Total Floor Area: 95,030 m²
Total Height/Floor: 160 m/ 34 F
Structure: Frame-Tube Structure

项目规划用地大致为矩形。我们采用简洁而有力度的两栋高层塔楼，通过裙房连为一体，最大限度地在地块内部形成整体的建筑群体感，从而形成标志性建筑。两栋塔楼一主一辅的布局形态，更好地体现了建筑的最佳高宽比。主体立面采用玻璃幕墙与垂直石材幕墙线条相结合的手法，外凸的线条强化了向上的动势，我们摒弃了花哨的造型和材料组合，而是通过严谨的比例推敲、细腻的材料组合，希望能使建筑在时代的沿革中具有更好的耐久性。

This project is located on an approximately rectangular site. Two simple but elegant high-rise tower buildings are connected through podium building into a whole, producing the most integral building image and creating a landmark. One of these two buildings is planned as main building and the other one as auxiliary building, so as to realize better height-to-width ratio of the building. The main facade is decorated with glass curtain wall and vertical stone curtain wall and uses protruded profile to realize upsoaring image. There is no sophisticated shape and material combination, but strict proportion and skillful material combination are adopted to produce better durability.

宁波银行象山大厦
Bank of Ningbo Xiangshan Building

建 设 单 位：宁波银行股份有限公司
功 能 用 途：金融、办公
设计/竣工年份：2008年/~
建 设 地 点：浙江省宁海县气象北路以西
总 建 筑 面 积：39 000 m²
总建筑高度/层数：80 m/21 F
结 构 形 式：框筒结构

Owner: Bank of Ningbo Co., Ltd.
Function: Finance, Office
Design/Completion Year: 2008/~
Construction Site: West of Qixiang North Road, Ninghai County, Zhejiang
Total Floor Area: 39,000 m²
Total Height/Floor: 80 m/ 21 F
Structure: Frame-Tube Structure

项目由三部分组成：地下两层，为必需的设备用房、汽车库（战时人防区域）；四层裙房，为银行的营业厅及办公用房，沿基地东侧布置；21层商务办公楼，主要为租售办公空间，位于基地西北角。建筑造型结合道路交叉口景观效果，主楼与裙房均为退台设计，减弱对城市空间的压迫感，丰富高层轮廓线。建筑整体以竖向线条为主，建筑材质主要为灰色LOW-E中空玻璃与深灰色石材，简洁明快，与金融业建筑形象相匹配。

The project is composed of three parts, including two basement floors which are used for necessary equipment room and parking lot (aerial defense basement in wartime), 4-floor podium building which is used as a bank service hall and offices and which is planned on the eastern part of the site, 21-floor business building which provides office spaces for rent and which is positioned at the northwest corner of the site. The building shape could introduce the crossroad landscape into the building space, and recessed design of main building and podium building alleviates compression of urban space and enriches the skyline. The building facades are mainly decorated with longitudinal lines and they are mainly made of grey LOW-E hollow glass and deep grey stone. Its simple and vivid image responses to the image of financial building.

杭州经济技术开发区水质检测、调度、控制中心综合楼
Water Quality Inspection, Dispatching and Control Center General Building in Hangzhou Economic and Technological Development Zone

建 设 单 位：杭州经济技术开发区资产经营集团有限公司
功 能 用 途：办公
设计/竣工年份：2010年/~
建 设 地 点：浙江省杭州市经济技术开发区
总 建 筑 面 积：48 600 m²
总建筑高度/层数：86.8 m/21 F
结 构 形 式：框剪结构

Owner: Hangzhou Economic and Technological Development Zone Assets Operation Group Co., Ltd.
Function: Office
Design/Completion Year: 2010/~
Construction Site: Economic and Technological Development Zone, Hangzhou City, Zhejiang
Total Floor Area: 48,600 m²
Total Height/Floor: 86.8 m/ 21 F
Structure: Frame-Shear Wall Structure

项目位于杭州市经济技术开发区新城内，金沙湖南岸。在建筑主体的平面选择上，我们尽量增大南北朝向的建筑体量，综合楼以小进深双面采光的板式建筑为主体，围合成"U"字形的建筑平面，中间为连续十多层高的中庭空间，朝七格北路一侧则每四层布置一个两层高的生态绿化平台。这种布局很好地体现了低碳、节能的设计思路，春秋两季可自然通风，自然采光让大量办公区可不用人工照明而节电；空中绿化平台则为办公人员提供了很好的休憩空间，创造出花园式的办公环境，体现了现代人文精神。

The project is located in the Economic and Technological Development Zone New Town in Hangzhou City and by the south bank of Jinsha Lake. During determination of plane layout of the building, best efforts are made to realize maximum building spaces facing the south and the north. The main part of the general building is a low-rise building which has small depth and double lighting facades and which encloses a U-shaped plane layout. An atrium more than 10-floor high is planned at the center and one 2-floor ecological green platform is designed on every four floors on the side facing Qige North Road. This layout well expresses its low-carbon and energy-saving concept and could realize natural ventilation in spring and autumn. Natural lighting could reduce power consumption in many office blocks; the overhead green platforms provide comfortable rest space for workers, creating garden-type office environment and embodying modern humanism spirit.

酒店建筑
Hotel Building

118	天台蓝海酒店	Tiantai Lanhai Hotel
119	台州方远大饭店	Taizhou Fangyuan Hotel
120	海宁长安大酒店	Haining Chang'an Grand Hotel
121	海口七星级产权式国际大酒店	Haikou Seven-star Condo International Hotel
122	浙江国际影视中心综合服务大楼	Zhejiang International Movie & Television Center Comprehensive Service Building
123	杭州西湖国宾馆后勤房	The Rear Service Room of Hangzhou West Lake State Guest House
124	湖北省农村信用社联合社培训中心	Training Center of Hubei Province Rural Credit Cooperatives Union
126	金都海洋公园配套服务中心	Supporting Service Center of Jindu Haiyang Park
127	义乌日信国际大酒店	Yiwu Rixin International Hotel
128	中大·西郊半岛富春希尔顿酒店	Zhongda·Xijiao Peninsula Fuchun Hilton Hotel
130	东方君悦	Grand Oriental Hyatt Hotel
131	横店国贸大厦会议中心	Conference Center of Hengdian International Trade Building
132	天台山温泉度假山庄改扩建项目	Renovation and Extension Project of Tiantai Mountain Hot Spring Resort
134	宿迁威尼斯度假酒店	Suqian Venice Resort Hotel
136	绍兴柯桥嘉悦广场	Shaoxing Keqiao Jiayue Square
137	常州凯纳商务广场	Changzhou Kaina Business Square
138	千岛湖润和度假酒店	Qiandao Lake Runhe Resort Hotel

天台蓝海酒店
Tiantai Lanhai Hotel

建设单位：浙江天台蓝海酒店有限公司
功能用途：五星级酒店
设计/竣工年份：2013年/~
建设地点：浙江省台州市天台县
总建筑面积：45 950 m²
总建筑高度/层数：63.9 m/16 F
结构形式：框架结构

Owner: Zhejiang Tiantai Lanhai Hotel Co., Ltd.
Function: Five-star Hotel
Design/Completion Year: 2013/~
Construction Site: Tiantai County, Taizhou City, Zhejiang
Total Floor Area: 45,950 m²
Total Height/Floor: 63.9 m/16 F
Structure: Frame Structure

项目设计地上主要功能为五星级品牌酒店，地下室为酒店后勤用房、配套设备机房、机动车库、非机动车库。酒店定位为天台县第一个五星级城市商务酒店。建筑立面采用石材框架为主体结构，局部辅以玻璃幕墙，整体保持一种协调感。建筑细部的处理采用竖向线条与实体相结合的手法，使得立面的比例达到最佳效果，又透射出深厚的文化底蕴。丰富的细部表达、各部分的比例与尺度的推敲，充分反映了五星级酒店的高贵与典雅。

The main function of the project aboveground is designed as five-star brand hotel, and the basement is designed as rear service room, supporting equipment room and garages for motor vehicles and non-motor vehicles of the hotel. The hotel is positioned as the first five-star city business hotel in Tiantai County. The facade of the building adopts stone framework as major structure, supplemented by glass curtain wall partly, which maintains the harmonization of the entire building. Details of the building are treated by combining vertical lines with the entity, which makes the facade reach the optimal proportion and reflects profound cultural connotations. Abundant detail expression and weighing of proportion and size of each part make an adequate reflection of the nobility and elegance of this five-star hotel.

台州方远大饭店
Taizhou Fangyuan Hotel

建 设 单 位：方远建设集团房地产开发有限公司
功 能 用 途：酒店、餐饮、会议
设计/竣工年份：2013年/~
建 设 地 点：浙江省台州市经济开发区
总建筑面积：63 311 m²
总建筑高度/层数：58.9 m/13 F
结 构 形 式：框剪结构

Owner: Fangyuan Construction Group Real Estate Development Co., Ltd.
Function: Hotel, Restaurant, Conference
Design/Completion Year: 2013/~
Construction Site: Economic Development Zone, Taizhou City, Zhejiang
Total Floor Area: 63,311 m²
Total Height/Floor: 58.9 m/13 F
Structure: Frame-Shear Wall Structure

项目位于台州市经济开发区内，功能为五星级酒店及其配套设施。建筑风格借鉴了欧式构图元素，结合现代材料做法，力求创造出独特的建筑形象，体现酒店的豪华与韵味。酒店主楼沿南侧主干道展开，东、西、北三面裙房围合形成内院，内院部分下沉，结合周边多层环廊的设置，构成了开敞大气的内部空间形象。

The project is located in Economic Development Zone in Taizhou City. The function is five-star hotel and its supporting facilities. The architectural style learns from European composition elements, combining with the practice of modern materials, striving to create unique architectural image and reflect the luxury and connotations of the hotel. The main building of the hotel extends along the main road at the south side and the skirt buildings at the east side, the west side and the north side form a garth, which is partly sinking. Open and magnificent interior space image is formed combining the surrounding multilayer corridors.

海宁长安大酒店
Haining Chang'an Grand Hotel

建设单位：海宁银桥置业有限公司
功能用途：酒店
设计/竣工年份：2012年/~
建设地点：浙江省海宁市长安镇开元路北侧
总建筑面积：60 306 m²
总建筑高度/层数：82.2 m/19 F
结构形式：框剪结构

Owner: Haining Yinqiao Real Estate Co., Ltd.
Function: Hotel
Design/Completion Year: 2012/~
Construction Site: North side of Kaiyuan Road, Chang'an Town, Haining City, Zhejiang
Total Floor Area: 60,306 m²
Total Height/Floor: 82.2 m/19 F
Structure: Frame-Shear Wall Structure

项目位于海宁市长安镇开元路北侧，修川路西侧，总用地面积23 434 m²，地形呈长条形，地势平坦。地块的西北侧是景观河，南侧和东侧是城市干道。酒店主体为19层高的板式客房，10层高的客房为附楼，裙房高4层，局部3层。客房主体布置在用地北侧，裙房一层用于配套商业和餐饮，二层用于商业及餐饮包厢，三层用于宴会、会议，四层用于娱乐，主楼和附楼为酒店客房。用地南侧和西北侧布置绿化广场，作为人流疏散广场使用，同时通过一层架空的方法将南北景观联系起来。

The project is located at the north side of Kaiyuan Road and west side of Xiuchuan Road, Chang'an Town, Haining City, and the total land area is 23,434 m². Landform is elongated and flat. Landscape river is located at the northwest side of the land and urban trunk roads are at the south side and east side. The hotel includes a 19-storey main building of slab-type guest rooms, a 10-storey annex of guest rooms and a 4-storey (partly 3-storey) skirt building. The guest rooms are mainly placed at the north side of the land. 1F of the skirt building is for supporting business and catering, 2F is for business and catering boxes, 3F is for banquet and conference and 4F is for entertainment. Main building and the annex are for guest rooms of the hotel. Green square is placed at the south side and northwest side of the land, used as square for evacuation of people. Meanwhile, the landscapes at the north and the south are connected by the empty space of 1F.

海口七星级产权式国际大酒店
Haikou Seven-star Condo International Hotel

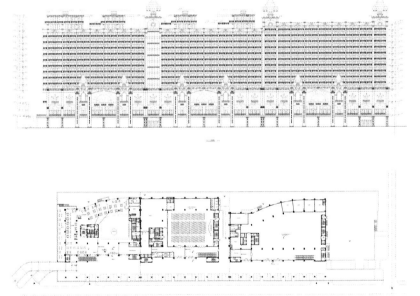

建设单位：海南尖峰置业有限公司
功能用途：酒店，书画创作、交易中心
设计/竣工年份：2013年/~
建设地点：海南省海口市
总建筑面积：116 000 m²
总建筑高度/层数：64.2 m/12+2 F
结构形式：框筒结构

Owner: Hainan Jianfeng Real Estate Co., Ltd.
Function: Hotel, Painting and Calligraphy Works Trade Center
Design/Completion Year: 2013/~
Construction Site: Haikou City, Hainan
Total Floor Area: 116,000 m²
Total Height/Floor: 64.2 m/12+ 2 F
Structure: Frame-Tube Structure

本项目由海南尖峰置业有限公司独资开发，该项目地处美丽的盈滨半岛国家AAAA级风景区内，与大海沙滩连成一体。项目按七星级产权式大酒店的标准定位设计，有1 200人以上同声传译会议大厅一个，1 000人左右同声传译会议大厅两个，能为规模1 200人以上的会议提供吃住服务；按照七星级高标准、高档次、高品位、高质量进行设计和施工。同时该项目也是全国第一个最大的集书画创作、交易于一体的大型交易市场，是国内最大的文化创作基地。

The project is developed by Hainan Jianfeng Real Estate Co., Ltd. with sole investment. The project is located at beautiful State AAAA Beauty Spot of Yingbin Peninsula, integrated with the sea and sand beach. The project is positioned and designed according to the standard of seven-star Condo hotel, including one conference hall for simultaneous interpretation with a capacity of more than 1,200 persons and two conference halls for simultaneous interpretation with a capacity of around 1,000 persons, which can provide dining and accommodation service for the conferences with the scale of more than 1,200 persons. It is designed and constructed according to seven-star high standard, high level, high grade and high quality. At the same time, it is the first largest trading market in China that integrates the creation and trading of painting and calligraphy. It is the largest cultural creation base in China.

浙江国际影视中心综合服务大楼
Zhejiang International Movie & Television Center Comprehensive Service Building

建设单位：浙江广播电视集团
功能用途：影视培训、住宿、餐饮
设计/竣工年份：2011年/~
建设地点：浙江省杭州市萧山区
总建筑面积：108 720 m²
总建筑高度/层数：126.45 m/29 F
结构形式：框筒结构

Owner: Zhejiang Radio and TV Group
Function: Film and TV Training, Hotel, Restaurant
Design/Completion Year: 2011/~
Construction Site: Xiaoshan District, Hangzhou City, Zhejiang
Total Floor Area: 108,720 m²
Total Height/Floor: 126.45 m/ 29 F
Structure: Frame-Tube Structure

项目由影视文化休闲娱乐用房、影视文化商务服务用房及相关辅助设施组成，两栋塔楼分别为29层和21层，高度分别约为130 m和80 m。两幢功能相近的大楼被规划为一个整体，以高档餐饮、豪华娱乐中心为主题的裙房把两幢大楼紧密连接，共同构筑成辐射整个周边地区的商务娱乐中心。这里既是为国际影视中心提供会议、培训、购物、餐饮、住宿等服务的综合配套区，又可以通过充分的市场开发，为国际影视中心各功能区域夯实经济基础，充分体现未来生活实践的最佳要素，成为园区创意者的快乐家园。

The project is composed of movie & television culture and entertainment room, movie & television culture and commercial service room and relevant auxiliary facilities; two tower buildings are respectively 29 floors and 21 floors, and the height is about 130 m or 80 m. Two buildings with similar functions are programmed as a whole. The podium building for top-grade catering and luxurious entertainment center connects the two buildings closely. The three buildings form the commerce and entertainment center covering the entire surrounding regions. It provides comprehensive supporting service such as conference, training, shopping, catering and accommodation for International Movie & Television Center. It also lays a solid economic foundation for each functional region of International Movie & Television Center by full development of market. It reflects the best elements of life practice in the future and becomes a happy homeland for the creators there.

杭州西湖国宾馆后勤房
The Rear Service Room of Hangzhou West Lake State Guest House

建设单位：杭州西湖国宾馆
功能用途：宾馆、会议
设计/竣工年份：2012年/~
建设地点：浙江省杭州市杭州西湖国宾馆
总建筑面积：1 450 m²
总建筑高度/层数：6.5 m/1 F
结构形式：框架结构

Owner: Hangzhou West Lake State Guest House
Function: Hotel, Conference
Design/Completion Year: 2012/~
Construction Site: Hangzhou West Lake State Guest House, Hangzhou City, Zhejiang
Total Floor Area: 1,450 m²
Total Height/Floor: 6.5 m/1 F
Structure: Frame Structure

项目总体布局采用江南园林式院落，复建部分与原有七号楼围合在一起。建筑尽量远离西湖湖面，与湖岸线距离大于85 m，保留并避开七号楼周边现有的珍贵树木，使建筑本身更好地隐蔽在山脚下，遮掩于青山绿树之间。建筑造型风格原则上沿用20世纪50年代起戴念慈先生的定式，屋顶以歇山顶为主，平坡结合。坡顶用黛色筒瓦，墙面以灰色面砖为主。亭台楼阁、假山奇石，恢复湖山胜景，园内文脉渊源，园外山环水抱，以求把该区域建成与西湖环境景观和谐统一的精致院落。

General layout of the project adopts garden-style courtyard of Jiangnan. The reconstructed part and original No. 7 Building form a circle. The building will be far from West Lake. The distance to the lake shoreline is more than 85 m. The project shall reserve and keep away from the existing precious trees around No.7 Building, which will hide the building at the foot of the mountain, covered by green mountains and trees. The architectural style follows the set pattern of Mr. Dai Nianci since the 1950s. The roof is irimoya-based, combining with flat slope. Top of slope uses greenish-black imbrex and the wall uses grey bricks. Pavilions, terraces and open halls, rockery and rare stones are made to recover wonderful scenery of lake and hills. The garden has context source inside and is surrounded by mountains and girdled by rivers outside. We strive to build the region to a delicate courtyard that is coordinated with the environmental landscape of West Lake.

湖北省农村信用社联合社培训中心
Training Center of Hubei Province Rural Credit Cooperatives Union

项目距离鄂州市主城区约24千米，交通方便，地理位置优越，有良好的湖景衬托，自然条件极其优越。项目由培训中心、室内运动馆、贵宾楼、后勤辅助用房、酒店五个功能区块组成。在规划设计中，设计师将不同类型的功能空间通过自由的交通路线组成有机的整体。培训中心作为建筑主体，布置在地块中偏西北位置，呈舒展曲线形，形成前后开放的空间。酒店布置在地块西北角，后勤辅助用房及室内运动馆布置在地块东北侧，贵宾楼布置在基地西南侧。整个项目区块，建筑的规划布局和风格与周边环境相融合，构筑成红莲湖地区城市景观的一部分。

The project is about 24 km away from main urban area of Ezhou, with convenient transportation, superior geographical location and beautiful lake view, natural conditions of which are extremely superior. The project is composed of five functional blocks—training center, indoor gym, VIP building, rear service auxiliary room and hotel. Functional spaces of different types become an organic whole by free traffic routes in planning and design. The training center is the main part of the building, which stands at the middle by northwest of the land plot, stretching in curvy shape and forming open space in the front and the back. The hotel is placed at the northwest corner of the land plot. The rear service auxiliary room and indoor gym are arranged at the northeast side of the land plot. VIP building is located at the southwest side of the base. The blocks of the whole project, the planning layout and style of the buildings coordinate with the surrounding environment, which becomes part of city landscape of Red Lotus Lake region.

建 设 单 位：航宇开发公司
功 能 用 途：培训中心、酒店
设计/竣工年份：2011年/~
建 设 地 点：湖北省鄂州市
总 建 筑 面 积：177 717 m²
总建筑高度/层数：120 m/32 F
结 构 形 式：框架结构
合 作 单 位：中国美术学院风景建筑设计研究院

Owner: Hangyu Development Corporation
Function: Training Center, Hotel
Design/Completion Year: 2011/~
Construction Site: Ezhou City, Hubei
Total Floor Area: 177,717 m²
Total Height/Floor: 120 m/32 F
Structure: Frame Structure
Cooperation Unit: The Design Institute of Landscape and Architecture, China Academy of Art

金都海洋公园配套服务中心
Supporting Service Center of Jindu Haiyang Park

建 设 单 位：金都房产集团有限公司
功 能 用 途：海洋馆配套用房
设计/竣工年份：2011年/~
建 设 地 点：浙江省杭州市
总 建 筑 面 积：52 291 m²
总建筑高度/层数：43.2 m/10 F
结 构 形 式：框架结构
合 作 单 位：杭州原田华建筑景观创意咨询有限公司

Owner: Jindu Real Estate Group Co., Ltd.
Function: Ocean Hall Supporting Facility
Design/Completion Year: 2011/~
Construction Site: Hangzhou City, Zhejiang
Total Floor Area: 52,291 m²
Total Height/Floor: 43.2 m/10 F
Structure: Frame Structure
Cooperation Unit: Hangzhou Yuantianhua Architectural Landscape Creative Planning Consultation Co., Ltd.

项目是金都海洋公园的二期工程，是整个地块内唯一的一幢高层建筑。它的造型既要与整个海洋馆建筑群融合，又要体现自己的个性。由于建筑基地位于钱塘江边，我们把建筑设计成游轮形状。整个建筑平面成"S"形摆布，从四层开始，层层退台，既化解了大体量建筑的厚重感，也增加了观景平台。顶层的飞檐增加了建筑的灵动感。建筑的外立面强调简洁、轻盈、通透。层层外挑的环廊增加了立面层次。建筑外装用材以铝板和玻璃为主，底材局部配以石材，增加建筑的稳重感。

The project is the 2ⁿᵈ phase project of Jindu Haiyang Park. It is the only high-rise building on the entire land plot. Its image must both coordinate with the entire buildings group of Aquarium and have its own independent personality. Because the building site is located beside the Qiantang River, we design the building into the shape like a cruise ship. The whole building has "S-shaped" plane layout, with backward terrace storey by storey from the 4ᵗʰ floor, which eliminates the heaviness of large building as well as increasing viewing platforms. The overhanging eaves of top floor make the building more alive. The facades of the building stress concision, lightness and transparency. The out-extended storeys of corridors strengthen the facade layering. Aluminium plates and glass are mainly used for exterior decoration of the building. Stones are used at part of the bottom to enhance the stable feeling of the building.

义乌日信国际大酒店
Yiwu Rixin International Hotel

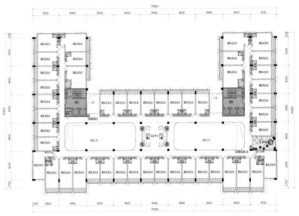

奇数层平面图

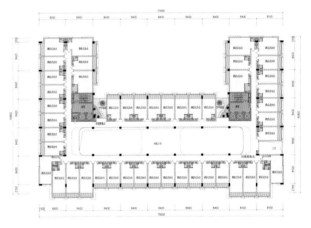

偶数层平面图

建设单位：日信置业有限公司
功能用途：酒店
设计/竣工年份：2011年/~
建设地点：浙江省义乌市廿三里街道
总建筑面积：42 162 m²
总建筑高度/层数：72 m/17 F
结构形式：框剪结构

Owner: Rixin Real Estate Co., Ltd.
Function: Hotel
Design/Completion Year: 2011/~
Construction Site: Niansanli Sub-district, Yiwu City, Zhejiang
Total Floor Area: 42,162 m²
Total Height/Floor: 72 m/17 F
Structure: Frame-Shear Wall Structure

项目位于义乌市商贸核心区域范围内的廿三里街道，商贸大道以北，开元街以西，主要建设内容为宾馆主体及附属配套设施。立面造型设计上，追求与义乌市商贸核心区域整个风貌相匹配，"大气、精致、和谐"，同时也注重其与周边已建或者在建高层、多层建筑的呼应和对话。外墙采用米色花岗岩干挂，结合通透的玻璃幕墙和竖向铝合金分隔线条，力求体现稳重挺拔的建筑特色。建筑整体追求利落大气的现代感，屋顶和局部位置采用了简欧风格的线条处理方式。

The project is located on the Niansanli Sub-district in the core commercial area of Yiwu City. The land plot is located at the north of Shangmao Avenue and the west of Kaiyuan Street. Main construction content includes main body of the hotel as well as the supporting facilities. It pursues "magnificence, delicacy and harmony" in the design of facade image, which matches the whole style and features of commercial core area of Yiwu City. At the same time, it also lays emphasis on the echo with the surrounding high-rise or multi-storey buildings already built or under construction. The external wall is covered with dry-hang cream-colored granites, combining with transparent glass curtain wall and vertical aluminium alloy separation lines, striving to reflect the steady and straight architectural features. The entire building pursues neat and magnificent modern look. The roof and some parts adopt lines of Jane European style.

中大·西郊半岛富春希尔顿酒店
Zhongda · Xijiao Peninsula Fuchun Hilton Hotel

会展中心东立面

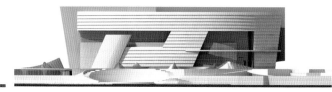

会展中心南立面

项目位于富阳市老城东侧，北面依山，南临富春江，拥有良好的自然与人文环境。项目用地面积 21 576 m²，地上建筑面积 51 013 m²，地下建筑面积 27 142 m²。它是希尔顿酒店旗下杭州市第一家定位为商务会议、休闲度假的超五星级豪华酒店，也是浙江省乃至整个长三角地区高端商务会议的首选酒店品牌。本项目将"帆"作为建筑形体的主题，沿江展开，形成千帆竞发的环境意象，努力创造一种自然、自信的扬帆起航的姿态，与富春江共生。

The project is located at the east side of old town, Fuyang City, with mountains in the north and Fuchun River to the south, having good natural and cultural environment. The site area of the project is 21,576 m², the aboveground floor area is 51,013 m² and the underground floor area is 27,142 m². It is the first super five-star luxury hotel for business conference and holidays subordinated to Hilton Hotel in Hangzhou City. It will become the first choice brand of hotel for high-end business conferences in the Zhejiang province or even the entire Yangtze River Delta area after it is completed. The project takes "sail" as theme of the architectural form, stretching along the river to form the environment image of racing of thousands of sails, and to create a natural and confident posture of setting sail, living together with Fuchun River.

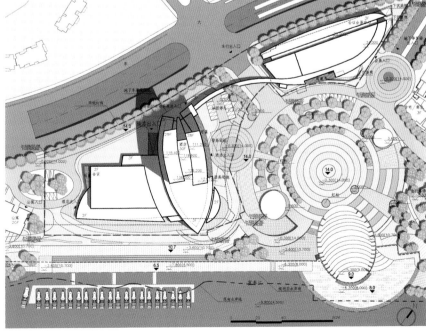

建设单位：富阳中大房地产有限公司
功能用途：酒店
设计/竣工年份：2011年/~
建设地点：浙江省富阳市
总建筑面积：78 155 m²
总建筑高度/层数：125.1 m/30 F
结构形式：框筒结构
合作单位：美国JWDA（上海骏地）

Owner: Fuyang Zhongda Real Estate Co., Ltd.
Function: Hotel
Design/Completion Year: 2011/~
Construction Site: Fuyang City, Zhejiang
Total Floor Area: 78,155 m²
Total Height/Floor: 125.1 m/30 F
Structure: Frame-Tube Structure
Cooperation Unit: America JWDA(Shanghai)

东方君悦
Grand Hyatt Hotel

建设单位：浙江全景置业有限公司
功能用途：酒店式公寓
设计/竣工年份：2008年/2013年
建设地点：浙江省杭州市钱江新城
总建筑面积：127 567 m²
总建筑高度/层数：129.7 m/34 F
结构形式：框筒结构

Owner: Zhejiang Quanjing Real Estate Co., Ltd.
Function: Seviced Apartment
Design/Completion Year: 2008/2013
Construction Site: Qianjiang New Town, Hangzhou City, Zhejiang
Total Floor Area: 127,567 m²
Total Height/Floor: 129.7 m/ 34 F
Structure: Frame-Tube Structure

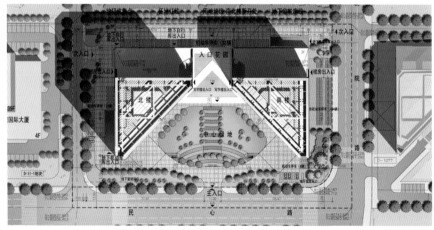

项目位于钱江新城民心路与剧院路交会处以北。整体建筑由A、B两座呈三角形的塔楼及与其相连接的三层裙房组成。A座高34层，B座高26层。裙房以商业金融类为主，四层以上为主楼，以写字间为主。外立面装修为玻璃幕墙和竖向铝合金装饰条，以加强建筑造型的挺拔感。为了让整个外立面更富神韵，在玻璃上印刷无机釉料，勾勒出一些几何形状的方格，从而演变成独有的方格彩釉玻璃幕墙。项目在钱江新城的超高层建筑群体中显得独一无二，尤为引人注目。

The project is located at the north of intersection of Minxin Road and Juyuan Road, Qianjiang New Town. The project is composed of two triangle tower buildings (A and B) and the 3-storey podium building connected with them. Tower A has 34 floors and Tower B has 26 floors. The podium building is mainly for commercial finance; floors above the 4th floor are main building, mainly for offices. The facades are decorated with glass curtain wall, with vertical aluminium alloy trim strips to make the architectural image taller and straighter. To make the facades more charming, the glass is printed with inorganic glaze and drawn some geometric square lattices to form unique square-lattice colored glazing glass curtain wall. It is unique in super high-rise buildings group in Qianjiang New Town, and very compelling.

横店国贸大厦会议中心
Conference Center of Hengdian International Trade Building

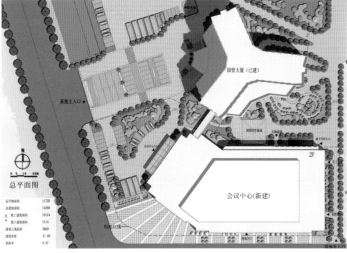

建设单位：横店影视城有限公司
功能用途：会议、餐饮
设计/竣工年份：2010年/~
建设地点：浙江省东阳市横店镇
总建筑面积：16 016 m²
总建筑高度/层数：17 m/2 F
结构形式：框架结构

Owner: Hengdian Film and TV City Co., Ltd.
Function: Conference, Restaurant
Design/Completion Year: 2010/~
Construction Site: Hengdian Town, Dongyang City, Zhejiang
Total Floor Area: 16,016 m²
Total Height/Floor: 17 m/ 2 F
Structure: Frame Structure

　　项目位于横店国贸大酒店用地内部，西侧为基地主入口，北侧为已建成的横店国贸大厦主楼。会议中心的主入口设置在建筑的西侧，正对主入口设置集中绿化场地，优化环境品质。在用地南侧沿府前路布置一条商业步行道路，并在城市道路转角处将空间放大，形成商业广场，以期形成良好的商业氛围。建筑体形方正简洁，局部有形体变化，立面简炼纯净。通过玻璃幕墙与石材墙面的合理搭配、线条与面的有机结合以及体量上的虚实对比，创造一个简约和谐、极富现代建筑美感的会议中心形象。

The project is located inside the land of Hengdian International Trade Hotel. Main entrance of the base is at the west side and main building of Hengdian International Trade Building is built at the north side. Main entrance of the conference center is set at the west side of the building, and green land is set directly facing the main entrance to optimize environment quality. One business pedestrian street is arranged along Fuqian Road at the south side of the land, and the space is amplified at the corner of the city road to form business square for good commercial atmosphere. The architectural shape is square and simple, with shape changes in some parts. The facade is clean and pure. Reasonable collocation of glass curtain wall and stone wall, organic combination of lines and planes as well as virtual-real comparison of space volume create a simple and harmonious image of conference center full of sense of beauty of modern architecture.

天台山温泉度假山庄改扩建项目
Renovation and Extension Project of Tiantai Mountain Hot Spring Resort

天台山以"佛宗道源,山水神秀"而誉满中外,佛、道文化源远流长。佛教的"天台宗"和道教的"南宗"都创于天台山,在历史上有着重要的地位和影响。项目位于天台县五峰桥,国赤路北侧,东临天然溪水,是原天皇药业厂房用地,紧临著名佛教寺庙国清寺。项目集合传统文化、佛道文化、温泉文化等多方面因素,采用唐代风格、古朴大方的建筑形式,创造出一处高品位、高档次之传世佳作。项目建成后将成为一家参照国家旅游局颁布的《星级饭店评定标准》之规定而设计的五星级酒店,功能齐全、设施先进,环境优美,是每一位客人会客、休憩、娱乐、收拾心情、整理思绪的好地方。

Tiantai Mountain is famous for its "origin of the Buddhism and Taoism, marvelous landscapes" at home and abroad, having a long history of Buddhism and Taoism culture. "Tiantai Sect" of Buddhism and "Nan Chung" of Taoism are created at Tiantai Mountain, which have important position and impact in the history. The project is located at Wufengqiao, Tiantai County, at the north side of Guochi Road and the west side of natural creek. The factory building of former Tianhuang Pharmaceutical company was there. It is next to the famous Buddhist temple—Guoqing Temple. It is an excellent work handed down from ancient times of good taste and high grade combining several factors like traditional culture, Buddhism and Taoism culture and hot spring culture, supplemented by the ancient but generous architectural form of Tang Dynasty. It will become a five-star hotel designed with reference to the stipulations of "Evaluation Standard of Star-rated Hotel" promulgated by National Tourism Administration after it is completed. It has complete functions, advanced facilities and beautiful environment. It will be a good place for every guest to receive visitors, to have a rest, to have entertainments, to tidy up the mood and to clear up the thoughts.

建设单位：天台山温泉度假山庄
功能用途：五星级酒店
设计/竣工年份：2010年/2013年
建设地点：浙江省天台县五峰桥，国赤路北侧
总建筑面积：48 788 m²
总建筑高度/层数：27.9 m/4 F
结构形式：框架结构

Owner: Tiantai Mountain Hot Springs Resort
Function: Five-star Hotel
Design/Completion Year: 2010/2013
Construction Site: North side of Guochi Road, Wufengqiao, Tiantai County, Zhejiang
Total Floor Area: 48,788 m²
Total Height/Floor: 27.9 m/ 4 F
Structure: Frame Structure

宿迁威尼斯度假酒店
Suqian Venice Resort Hotel

建 设 单 位：江苏亿泰旅业发展有限公司
功 能 用 途：酒店
设计/竣工年份：2009年/2012年
建 设 地 点：江苏省宿迁市
总建筑面积：50 000 m²
总建筑高度/层数：15.6 m/4 F
结 构 形 式：框架结构

Owner: Jiangsu Yitai Tourism Development Co., Ltd.
Function: Hotel
Design/Completion Year: 2009/2012
Construction Site: Suqian City, Jiangsu
Total Floor Area: 50,000 m²
Total Height/Floor: 15.6 m/ 4 F
Structure: Frame Structure

项目位于宿迁市骆马湖边的湖滨新城，工程由一个Best Western威尼斯度假酒店、一个经济型酒店和钟楼组成。将威尼斯度假酒店布置在临湖的西侧，使最多的客房拥有湖景；将行政客房布置在相对僻静的临内湖的一侧，综合了景观和私密性要求。会议厅和宴会厅靠近客房，方便使用。餐厅、SPA、康乐三部分都兼有对外功能，可以独立运营。景观设计充分利用湖景、湖水、夕阳、温泉及风景区环境资源，建构威尼斯水城风貌。

The project is located at Hubin New Town beside Luoma Lake, Suqian City. The project is composed of one Best Western Venice Resort Hotel, one economy hotel and bell tower. Venice Resort Hotel stands at the west lakeside, which enables the most guest rooms with lake view; executive rooms are arranged at quiet side near internal lake, which meets the requirements of landscape and privacy. Conference hall and banquet hall are near the guest rooms, convenient for using. Opened restaurant, SPA, recreation & entertainment are opened to the public, which can be operated independently. It makes full use of lake view, lake water, setting sun, hot spring and environmental resources of beauty spot in landscape design to construct the style and features of Venice Water City.

绍兴柯桥嘉悦广场
Shaoxing Keqiao Jiayue Square

建 设 单 位：浙江嘉悦实业有限公司
功 能 用 途：酒店、商业、住宅
设计/竣工年份：2009年/~
建 设 地 点：浙江省绍兴市柯桥
总 建 筑 面 积：104 300 m²
总建筑高度/层数：150 m/41 F
结 构 形 式：框筒结构
合 作 单 位：美国JWDA（上海骏地）

Owner: Zhejiang Jiayue Real Estate Co., Ltd.
Function: Hotel, Commerce, Residence
Design/Completion Year: 2009/~
Construction Site: Keqiao, Shaoxing City, Zhejiang
Total Floor Area: 104,300 m²
Total Height/Floor: 150 m/ 41 F
Structure: Frame-Tube Structure
Cooperation Unit: America JWDA(Shanghai)

本工程位于浙江省绍兴市柯桥轻纺城，邻近绍兴市政府，南临城市主干道群贤路，正南直面开阔大气、秀丽无边的瓜渚湖，静静流淌的娄直江在东面将其轻轻环抱。项目用地面积22 319 m²，地上建筑面积74 520 m²，地下建筑面积29 779 m²。整体建筑为希尔顿逸林酒店、精品名牌商业店、精装平层大宅及空中别墅共同构成的城市综合体。在垂直空间上，创造了集居住、商务、休闲、娱乐、消费于一体的城市高端生活方式。

The project is located at Keqiao Light Textile City, Shaoxing City, Zhejiang, near Shaoxing Municipal Government, to the north of city main road—Qunxian Road. It faces the broad, magnificent and elegant Guazhu Lake directly at the south, and the quiet Louzhi River gently embraces the project at the east side. The site area of the project is 22,319 m², the aboveground floor area is 74,520 m² and the underground floor area is 29,779 m². The entire architecture is an urban complex consisting of Hilton Double Tree Hotel, business of boutique brand shops, flat-layer house of refined decoration and penthouse apartment. It creates a high-end city life style by combining residence, business, relaxation, entertainment and consumption into one in vertical space.

常州凯纳商务广场
Changzhou Kaina Business Square

建 设 单 位：常州洛察纳房地产开发有限公司
功 能 用 途：酒店、酒店式公寓
设计/竣工年份：2008年/2010年
建 设 地 点：江苏省常州市
总 建 筑 面 积：106 416 m²
总建筑高度/层数：180 m/14 F
结 构 形 式：框剪结构

Owner: Changzhou Luochana Real Estate Development Co., Ltd.
Function: Hotel, Seviced Apartment
Design/Completion Year: 2008/2010
Construction Site: Changzhou City, Jiangsu
Total Floor Area: 106,416 m²
Total Height/Floor: 180 m/ 14 F
Structure: Frame-Shear Wall Structure

本项目包括两幢综合体建筑，南部01地块为集酒店、酒店式公寓、商业、餐饮等功能于一体的超高层综合建筑体，北部02地块为集停车、商业、居住等功能于一体的高层综合建筑体。我们将塔式超高层公寓布置于该地块南部，呈东西走向，争取最大南向日照。弧形板式四星级酒店位于地块东南部，东侧布置大面积的广场和景观绿化。裙房三层满铺，底层为商铺和错层式停车库。主楼为14层的高层板式住宅。建筑体量运用空间组织的穿插、渗透、收放，突出空间的变化，丰富整体空间效果，创造凹凸有致、错落自如的城市界面。

The project includes two building complexes. Land plot 01 at the south is a super high-rise building complex for hotel, serviced apartment, business, catering and other functions. Land plot 02 at the north is a high-rise building complex for parking, business, residence and other functions. We place the tower-type super high-rise apartment at the south of the land plot, which runs east-west for the most south sunlight. The arc slab-type four-star hotel is located at the southeast side of the land plot; there are a square and landscape greening of large area at the east side. The skirt building has three floors full of stores. The ground floor is for stores and split-level parking garages. Main building is a 14-storey high-rise slab-type residence. The space is interspersed, permeated and changed to enrich the entire space effect and to create a nice, well-arranged and free city interface.

千岛湖润和度假村酒店
Qiandao Lake Runhe Resort Hotel

该项目位于千岛湖东西向的半岛——梦姑岛上。该岛地形起伏较大，山坡陡峭。规划设计时充分考虑到设计的难度，尽量结合原自然山水条件和地形地貌，利用最佳景观来排布建筑物，使建筑与周围环境融为一体。由于酒店规模较大，在建筑平面和立面设计中既要减少对自然景观和环境的破坏，又要避免形成庞大的建筑体量，设计充分利用高低错落的地形，将项目化解为多个体量。建筑体形上采用局部退台的方式与山体走势有机结合，使整个度假酒店的建筑体量高高低低、错落有致，形成了比较丰富的建筑天际线，也丰富了建筑的形态。

The project is located at the east-west peninsula—Menggu Island in Qiandao Lake. The island has large terrain variations and steep hills. The design difficulty is fully considered in planning and design. We try to combine with original natural landscape conditions and landform, utilize the best landscape to arrange the buildings, so as to make the buildings coordinated with surrounding environment. Because the hotel is of large scale, we need to reduce the damages to the natural landscape and environment in plane and facade design of building as well as avoid huge space volume. We make full use of terrain of different heights in the design to resolve many space volumes. We make organic combination of backward terrace in some parts and trend of the mountain in architectural shape to make the entire resort hotel well-arranged with different space volume, to form abundant architectural skyline and to enrich architectural forms.

建设单位：杭州千岛湖润和度假村有限公司
功能用途：度假酒店
设计/竣工年份：2006年/2011年
建设地点：浙江省杭州市千岛湖梦姑岛
总建筑面积：53 000 m²
总建筑高度/层数：19.8 m/6 F
结构形式：框架结构

Owner: Hangzhou Qiandao Lake Runhe Resort Hotel Co., Ltd.
Function: Resort
Design/Completion Year: 2006/2011
Construction Site: Menggu Island, Qiandao Lake, Hangzhou City, Zhejiang
Total Floor Area: 53,000 m²
Total Height/Floor: 19.8 m/6 F
Structure: Frame Structure

商业建筑
Commercial Building

- 142 遵义国际商贸城
- 144 诸暨国际商贸城
- 145 北海东盟国际商贸城
- 146 重庆朝天门国际商贸城
- 148 大连辽渔国际水产品市场
- 149 五峰金东山家居建材城
- 150 江苏涟水2011—15号宗地项目
- 152 泰州泰茂城
- 153 英冠·天地城
- 154 湖北鄂州航宇国际商贸城
- 155 杭州转塘狮子村商业项目
- 156 杭州滨江76号A地块——世贸商务楼
- 157 钱江世纪城富丽大厦
- 158 杭州江口股份经济合作社等3家经合社商业综合用房
- 159 杭州方家畈股份经济合作社商业用房
- 160 昆明螺蛳湾国际商贸城
- 162 杭州厚仁商业街
- 163 宁波江北万达广场

- 142 Zunyi International Trade City
- 144 Zhuji International Trade City
- 145 Beihai ASEAN International Trade City
- 146 Chongqing Chaotianmen International Trade City
- 148 Dalian Liaoyu International Aquatic Product Market
- 149 Wufeng Jindongshan Home Furnishing and Building Materials City
- 150 Jiangsu Lianshui 2011-15 Parcel Project
- 152 Taizhou Taimao City
- 153 Yingguan • Tiandi City
- 154 Hubei Ezhou Hangyu International Trade City
- 155 Commercial Project of Shizi Village, Zhuantang Town, Hangzhou City
- 156 Lot A, No.76 Binjiang, Hangzhou – Shimao Business Building
- 157 Fuli Building of Qianjiang Century CBD
- 158 Comprehensive Commercial Buildings of 3 Economic Cooperatives Including Hangzhou Jiangkou Share Economic Cooperative
- 159 Commercial Buildings of Hangzhou Fangjiafan Share Economic Cooperative
- 160 Kunming Luoshiwan International Trade City
- 162 Hangzhou Houren Commercial Street
- 163 Ningbo Jiangbei Wanda Plaza

遵义国际商贸城
Zunyi International Trade City

项目地处贵州省遵义市红花岗经济开发区，总占地面积约73.7 hm²。分2~3期开发建设，其中第1期建筑面积约520000 m²。项目集商品展示贸易（场馆式市场、街式市场）、商务办公、金融服务、教育培训、新品推广发布、电子商务、酒店、SOHO公寓、住宅与旅游购物于一体，是集约化、规模化、国际化、现代化程度较高的大型综合性商贸集散中心。

Located in Honghuagang Economic Development Zone in Zunyi City, Guizhou Province, the project covers a site area of about 73.7 hectares. The project is constructed in 2 to 3 phases, of which, the floor area of the first phase is about 520,000 m². The project is integrated with commodity display trade (venue-type market and street-type market), business office, financial service, educational training, promotion and distribution of new products, e-commerce, hotel, SOHO apartment, residential house and tourism and shopping. It is an intensive, international, modern, and large comprehensive collecting and distribution trade center.

建设单位：义乌市思达投资有限公司	Owner: Yiwu Star Investment Co., Ltd.
功能用途：市场	Function: Market
设计/竣工年份：2011年/~	Design/Completion Year: 2011/~
建设地点：贵州省遵义市红花岗经济开发区	Construction Site: Honghuagang Economic Development Zone, Zunyi City, Guizhou
总建筑面积：1 600 000 m²	Total Floor Area: 1,600,000 m²
总建筑高度/层数：18-67 m/4-18 F	Total Height/Floor: 18-67 m/ 4-18 F
结构形式：框剪结构	Structure: Frame-Shear Wall Structure

诸暨国际商贸城
Zhuji International Trade City

建设单位：诸暨浣江国际商贸城开发有限公司
功能用途：市场
设计/竣工年份：2013年/~
建设地点：浙江省诸暨市
总建筑面积：416 661 m²
总建筑高度/层数：22.8 m/4 F
结构形式：框架结构

Owner: Zhuji Huanjiang International Trade City Development Co., Ltd.
Function: Market
Design/Completion Year: 2013/~
Construction Site: Zhuji City, Zhejiang
Total Floor Area: 416,661 m²
Total Height/Floor: 22.8 m/4 F
Structure: Frame Structure

项目位于诸暨市中心城区南部。一期项目位于商贸城北侧。基地建设用地面积约 24.6 hm²。本项目以块状市场布置，以标准化的市场模块为基本单元设计，可满足市场的弹性和灵活性需求，满足未来变动的需要。在建筑形象的处理上，本方案以诸暨当地特产"香榧"及"珍珠"为立面构思的出发点，通过立面构件的交叉、强壮有力的枝权状柱子、晶莹剔透的球状玻璃将诸暨的地方特色抽象化、建筑化，使整个诸暨商贸城的立面外观产生了强烈的视觉冲击力，具有极强的可识别性和标志性。

The project is located in the south of the downtown area of Zhuji City. The Phase I project is located in the north side of the trade city. The land for construction of the base is about 24.6 hm². The project is arranged with the block market structure and takes standard market modules as basic units for design, which can meet the elasticity and flexibility demand of the market and meet the demand in case of any change in the future. In the treatment of the building image, the special local products of Zhuji—"torreya" and "pearl" are taken as the starting point for facade design. The crossing of facade components, the strong branch-shaped columns, and the crystal clear spherical glass are abstract and architectural forms of the local characteristics of Zhuji, which gives the whole facade look a strong visual impact and makes it easy to identify.

北海东盟国际商贸城
Beihai ASEAN International Trade City

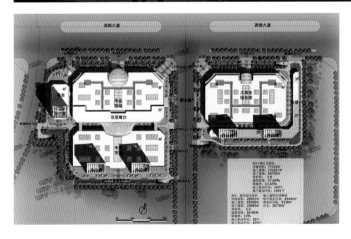

建设单位：北海东盟国际商贸城
功能用途：商业、办公
设计/竣工年份：2013年/~
建设地点：广西北海市海城区东部
总建筑面积：300 000 m^2
总建筑高度/层数：85 m/22 F
结构形式：框架结构

Owner: Beihai ASEAN International Trade City
Function: Commerce, Office
Design/Completion Year: 2013/~
Construction Site: Eastern Part of Haicheng District, Beihai City, Guangxi
Total Floor Area: 300,000 m^2
Total Height/Floor: 85 m/22 F
Structure: Frame Structure

项目基地位于北海市海城区东部，紧靠城市快速路西南大道与湖北路交叉口。周边交通便利，物流通畅。规划用地约7.77 hm^2。以超前的规划设计，将项目打造为一个立足北部湾、面向大西南、对接东南亚、服务全世界的现代化的商贸核心。建成后它将是集商品展示、交易、会展服务、物流配送、创意开发服务于一体，以流通为主体的一站式综合商品展示和交易平台。

The project is located in the east side of Haicheng District, Beihai City, next to the intersection of the city expressway, Southwest Avenue and Hubei Road. The project enjoys a convenient traffic, which makes the logistics expedite and unobstructed. The planning site area is about 7.77 hm^2. With advanced planning and design, it aims to create a modern trade center that is based on Beibu Gulf, faces the southwestern part of China, connects Southeast Asia and serves the whole world. After it is completed, the project will be a one-stop comprehensive commodity display and trade platform integrating commodity display, trade, exhibition service, logistics distribution and innovation and development service.

重庆朝天门国际商贸城
Chongqing Chaotianmen International Trade City

项目基地南起茶涪路，东至重庆绕城高速，北至快速路三横线，西临迎龙镇。规划总用地面积约170.9 hm²。项目是一个集批发市场、配套服务、物流、仓储、商务办公、生活配套以及会展、高端零售、高端酒店等的整体式多元功能体。设计通过点、线、面结合以及立体绿化的引入，形成商贸城良好的景观绿化空间。强调以南北向贯穿市场的商业内街轴线以及主入口正对的轴线为景观主轴，在两条主轴交叉处设置中心大型景观，作为市场主入口的对景以及市场的景观中心。

The project base starts from Chabei Road in the south, ends at Chongqing Ring Expressway in the east and Rapid Road Third Horizontal Line in the north, and adjoins Yinglong Town in the west. The total planning site area is about 170.9 hm². The project is a multi-functional entity integrating wholesale market, supporting service, logistics, warehousing, business office, life supporting facilities, exhibition, high-end retail and high-end hotel. In the design, an excellent landscape greening space is formed with the combination of point, line and face and the introduction of vertical planting. Designers take the inside commercial street axis running through the market in the north-south direction and the axis directly facing the main entrance as the principle axes of the landscape, set large central landscape at the crossing of the two principle axes, and take it as the opposite scenery of the main market entrance and the landscape center of the market.

建 设 单 位：重庆市朝天门国际商贸城有限公司
功 能 用 途：批发交易市场
设计/竣工年份：2011年/~
建 设 地 点：重庆市南岸区茶园组团
总建筑面积：5 500 000 m²
总建筑高度/层数：23.9 m/5 F
结构形式：框架结构

Owner: Chongqing Chaotianmen International Trade City Co., Ltd.
Function: Wholesale Market
Design/Completion Year: 2011/~
Construction Site: Chayuan Plot, Nan'an District, Chongqing City
Total Floor Area: 5,500,000 m²
Total Height/Floor: 23.9 m/5 F
Structure: Frame Structure

大连辽渔国际水产品市场
Dalian Liaoyu International Aquatic Product Market

建 设 单 位：辽宁省大连海洋渔业集团公司
功 能 用 途：市场
设计/竣工年份：2012年/~
建 设 地 点：辽宁省大连市大连湾以北
总 建 筑 面 积：119 600 m²
总建筑高度/层数：22.65 m/4 F
结 构 形 式：框架结构

Owner: Dalian Ocean Fishery Group of Corporations, Liaoning
Function: Market
Design/Completion Year: 2012/~
Construction Site: North of Dalian Bay, Dalian City, Liaoning
Total Floor Area: 119,600 m²
Total Height/Floor: 22.65 m/ 4 F
Structure: Frame Structure

项目地块南临黄海，位于大连湾以北，规划道路以南，在整个都市渔港RBD区域中占据桥头堡的位置。地块南北长约320 m，东西宽120 m至210 m不等，占地约5.29 hm²。项目今后的业态以水产品批发交易市场、综合市场、办公、餐饮为主。规划建筑主要由4幢多层的市场单元块组成。整个市场通过空中连廊相连接，形成一个完整的商业综合体。

The project lot adjoins the Yellow Sea in the south and is located in the north side of the Dalian Bay and the south side of the planning road. It is situated at the bridgehead of the RBD of the whole urban fishing port. The plot is about 320 m long in the north-south direction and about 120 m to 210 m wide in the east-west direction, covering a site area of about 5.29 hectares. The future business type of the project will be featured by wholesale trade market of aquatic products, comprehensive market, office and restaurants. The buildings are mainly composed of 4 multiple-floor market units. The whole market will form a complete business complex with space corridors.

五峰金东山家居建材城
Wufeng Jindongshan Home Furnishing and Building Materials City

建设单位：五峰金东山商业集团有限公司
功能用途：住宅、市场
设计/竣工年份：2012年/~
建设地点：湖北省宜昌市五峰县
总建筑面积：72 000 m²
总建筑高度/层数：57 m/19 F
结构形式：框架结构

Owner: Wufeng Jindongshan Commercial Group Co., Ltd.
Function: Residence, Market
Design/Completion Year: 2012/~
Construction Site: Wufeng County, Yichang City, Hubei
Total Floor Area: 72,000 m²
Total Height/Floor: 57 m/19 F
Structure: Frame Structure

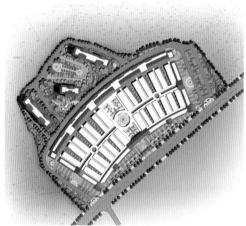

项目位于湖北省宜昌市五峰县东西路，占地面积约48 500 m²。项目拟打造成为以品牌家电、品牌家具及家居软装、建材销售、商务办公、居住及配套服务等为主体的综合体。在设计中，我们力图展现一组传统、简洁、大气开放的建筑形象，从而成为区域的形象走廊及区域发展的推动力。考虑到地块现状及周边的情况，将项目的设计划分为两个分区：市场区块和居住区块。在划分区块的时候坚持合理分区、资源共享的原则，使各区域既相对独立，又通过道路与空间有机地联系在一起。

The project is located in Dongxi Road, Wufeng County, Yichang City, Hubei, with a site area of about 48,500 m². The project is expected to be built into a complex featured by brand household appliances, brand furniture, home furnishing, sales of building materials, business office, residence and supporting services. In the design, we seek to present a traditional, simple, grand and open building image and make it a driving force for regional image corridor and regional development. Considering the actual situation of the lot and the surroundings, the project is divided into two zones for design: market zone and residence zone. We stick to the principles of reasonable zoning and resource sharing during the zoning, to make each zone independent from each other yet dynamically connected through roads and spaces.

江苏涟水2011—15号宗地项目
Jiangsu Lianshui 2011-15 Parcel Project

建 设 单 位：江苏金三源集团
功 能 用 途：商业、住宅
设计/竣工年份：2011年/~
建 设 地 点：江苏省淮安市涟水县
总 建 筑 面 积：666 800 m²
总建筑高度/层数：100 m/32 F
结 构 形 式：框架结构
合 作 单 位：中国美术学院风景建筑设计研究院

Owner: Jiangsu Jinsanyuan Group
Function: Commerce, Residence
Design/Completion Year: 2011/~
Construction Site: Lianshui County, Huaian City, Jiangsu
Total Floor Area: 666,800 m²
Total Height/Floor: 100 m/ 32 F
Structure: Frame Structure
Cooperation Unit: Design Institute of Landscape and Architecture China Academy of Art

　　本案位于江苏省涟水县，用地面积144 452 m²。在建筑整体的平面布局设计上，一层采用导向性明确、通达东西的商业街，二层采用退台形式，三层引入曲线元素，S形的商业界面贯穿东西，形成丰富的商业流线，同时与建筑内部的环境产生互动，构筑动人的空间效果。同时，景观广场利用中心商业街道串联，通过南北向的景观步道同城市外部联系，达到建筑与城市的空间对话，体现整个商业体系服务于城市的立意。

The project is located in Lianshui County, Jiangsu, with a site area of 144,452 m². In the overall plane layout design of the building, the first floor is a commercial street with clear orientation and connects the east and the west; the terraced form is adopted on the second floor; on the third floor, curve element is introduced. The S-shaped business interface connects the east and the west, forming a rich commercial streamline and at the same time interacting with the internal environment of the building and thus creating amazing spatial effects. Meanwhile, the landscape square is connected in series with the central commercial street and is connected with the outside through the landscape footpath in the north-south direction, which realizes spatial dialogue between the building and the city and shows the idea that the whole commercial system serves the city.

泰州泰茂城
Taizhou Taimao City

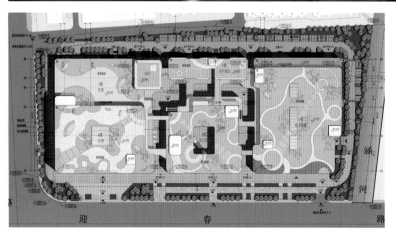

建设单位：江苏华辰房地产开发有限公司
功能用途：商业综合体
设计/竣工年份：2012年/～
建设地点：江苏省泰州市
总建筑面积：177 586 m²
总建筑高度/层数：48 m/9 F
结构形式：框架结构

Owner: Jiangsu Huachen Real Estate Development Co., Ltd.
Function: Commercial Complex
Design/Completion Year: 2012/~
Construction Site: Taizhou City, Jiangsu
Total Floor Area: 177,586 m²
Total Height/Floor: 48 m/ 9 F
Structure: Frame Structure

项目位于泰州市海陵区城河东侧，为城东核心地带，用地面积53 591 m²，建筑占地面积26 033 m²。它的发展将形成城东乃至泰州的一个新地标，对区域服务和完善城市规划有着极其重要的作用。本项目还在于营造一处开放的综合性购物休闲公园，包括高档百货公司、室内步行街区、休闲娱乐设施、延伸教育服务设施、IMAX影院、时尚电子商厦等。

Located in the east side of the City River in Hailing District, Taizhou City, the project is in the cental area of Chengdong Area, with a site area of 53,591 m² and a footprint of 26,033 m². The project is expected to become a new landmark in Chengdong Area and even in Taizhou City, which is of great importance to regional services and the improvement of the urban planning. The project is also expected to form an open comprehensive shopping and leisure park, including high-end department store, indoor walking street, leisure and recreational facilities, extension education service facilities, IMAX cinema, fashionable electronic building, etc.

英冠·天地城
Yingguan·Tiandi City

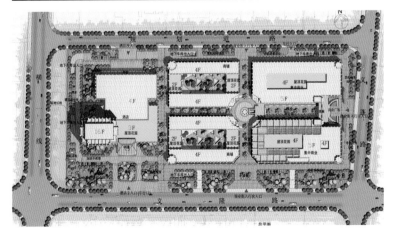

建 设 单 位：浙江中冠房地产开发有限公司
功 能 用 途：酒店、商业
设计/竣工年份：2012年/~
建 设 地 点：浙江省杭州市萧山区
总 建 筑 面 积：178 691 m²
总建筑高度/层数：64.4 m/17 F
结 构 形 式：框剪结构

Owner: Zhejiang Zhongguan Real Estate Development Co., Ltd.
Function: Hotel, Commerce
Design/Completion Year: 2012/~
Construction Site: Xiaoshan District, Hangzhou City, Zhejiang
Total Floor Area: 178,691 m²
Total Height/Floor: 64.4 m/17 F
Structure: Frame-Shear Wall Structure

本项目西侧地块定位为五星级酒店，东侧地块定位为围合式商铺以及大型集中商业体。为了获得较好的商业环境，设计采用了顺应地形的矩形布局。在对场地分析所形成的建筑布局中，我们获得了一个舒展的面宽。通过对规划条件中"限高66 m"的分析，我们对原先较为宽阔的体量进行了处理。而在建筑细部的处理上我们采用装饰艺术派（ART DECO）的建筑语汇，采用竖向线条与实体相结合的手法，使得立面的比例达到最佳的效果。

The west lot of the project is positioned as a five-star hotel, while the east lot, as enclosed shops and large concentrated business entity. To obtain a better business environment, a rectangular layout adapted to the landform is adopted. The architectural composition formed through the analysis of the site provides us with a stretching face width. Based on the "66 m max headroom" specified in the planning conditions, we have made adjustment to the original broad masses. While in the treatment of the details of the building, we have adopted ART DECO. The technique of combining the vertical lines with the entity brings the best of the proportion of the facade.

湖北鄂州航宇国际商贸城
Hubei Ezhou Hangyu International Trade City

建设单位：湖北航宇置业有限公司
功能用途：商场、酒店
设计/竣工年份：2010年/2012年
建设地点：湖北省鄂州市
总建筑面积：185 345 m²
总建筑高度/层数：50.5 m/13 F
结构形式：框架结构

Owner: Hubei Hangyu Real Estate Co., Ltd.
Function: Shopping Mall, Hotel
Design/Completion Year: 2010/2012
Construction Site: Ezhou City, Hubei
Total Floor Area: 185,345 m²
Total Height/Floor: 50.5 m/ 13 F
Structure: Frame Structure

本项目位于花湖经济开发区内。功能形态涵盖家具广场、建材市场、超市及娱乐区、快捷酒店、幼儿园及辅助配套用房。项目的设计划分为五个分区：月星家居广场、超市娱乐区、建材市场及酒店区、步行街和辅助配套区。各个区块在功能、定位以及面向的人群等方面有区别，所以在划分区块的时候坚持合理分区、资源共享的原则。使各区域既相对独立，又通过道路有机地联系在一起，从而符合本区块的城市脉络及项目本身的规划性质。

The project is located in Huahu Economic Development Zone. The project includes furniture market, building materials market, supermarket and entertainment zone, express hotel, kindergarten and auxiliary and supporting housing. The project is divided into five zones for design: Yuexing Home Furnishing Square, supermarket and entertainment zone, building materials market and hotel zone, walking street, and auxiliary and supporting zone. As the zones have their own functions, positioning and target customers, the principle of reasonable zoning and resources sharing is adopted during the zoning, which makes the zones relatively independent from each other yet dynamically connected through roads, and complies with the city context of the region and the planning nature of the project itself.

杭州转塘狮子村商业项目
Commercial Project of Shizi Village, Zhuantang Town, Hangzhou City

建设单位：杭州元山置业有限公司
功能用途：商场、办公、酒店
设计/竣工年份：2012年/~
建设地点：浙江省杭州市之江度假区
总建筑面积：106 233 m²
总建筑高度/层数：48 m/11 F
结构形式：板柱结构

Owner: Hangzhou Yuanshan Real Estate Co., Ltd.
Function: Shopping Mall, Office, Hotel
Design/Completion Year: 2012/~
Construction Site: Zhijiang Holiday Zone, Hangzhou City, Zhejiang
Total Floor Area: 106,233 m²
Total Height/Floor: 48 m/ 11 F
Structure: Slab-Column Structure

本项目建设用地面积为35 187 m²，建筑总占地面积为14 074 m²。根据前期分析，认为该地块作为旅游品商场及酒店的商业综合体较为合理，可以充分利用周边良好的旅游资源及交通条件，营造独具江南特色的商业形态。在设计上本建筑引用传统江南民居的布局形式，形成大型庭院。景观处理引入水乡民居特色，从建筑外部的景观水池到室内的水街，充分灵活地运用江南元素营造独具特色的商业氛围，另外以马头墙、坡屋顶等建筑形式加以点缀，赋予整个建筑独特的典雅、精致和人文神韵。

The construction land area of the project is 35,187 m²; the total footprint is 14,074 m². According to the preliminary analysis, it's believed that it is reasonable to build a commercial complex including shopping malls of travel products and hotels. The excellent tourism resources and traffic conditions in the surroundings may be made full use of to create a business form featured by the characteristics of regions south of the Yangtze River. In the design, the layout of traditional folk houses in regions south of the Yangtze River is adopted, forming a large courtyard. In the landscape treatment, the characteristics of the folk houses in regions south of the Yangtze River are introduced. From the landscape pond outside the building to the indoor water street, the elements of regions south of the Yangtze River are used to create a characteristic commercial atmosphere. Additionally, the building is decorated with escarpment walls, slope roofs and other architectural forms, which makes the whole building elegant, dedicate and cultural.

杭州滨江76号A地块——世茂商务楼
Lot A, No.76 Binjiang, Hangzhou-Shimao Business Building

建设单位：海墅房地产开发（杭州）有限公司
功能用途：商业、办公
设计/竣工年份：2012年/~
建设地点：浙江省杭州滨江区
总建筑面积：178 000m²
总建筑高度/层数：159.5m/ 35F
结构形式：框架结构

Owner: Haishu Real Estate Development (Hangzhou) Co., Ltd.
Function: Commerce, Office
Design/Completion Year: 2012/~
Construction Site: Binjiang District, Hangzhou City, Zhejiang
Total Floor Area: 178,000m²
Total Height/Floor: 159.5m/ 35F
Structure: Frame Structure

项目位于杭州市钱塘江南岸的滨江区，定位为5A甲级商务写字楼及配套商业用房，配备一定数量的休闲娱乐、餐饮、电影院等配套服务设施。建筑造型为现代风格，塔楼沿江面以两片挺直的玻璃幕墙隐喻风帆，沿住宅面以相互错位的窗和墙形成富有变化的肌理。建筑立面设计强调竖向线条分割，塑造出挺拔向上的建筑形象，整体风格干练大气。裙房主要采用高档的花岗石外墙，并与铝合金网格相结合，既彰显建筑的现代感，又体现其文化内涵。

The project is located in Binjiang District in the south bank of the Qiantang River in Hangzhou City. It is positioned as 5A Grade A office tower and supporting commercial buildings, equipped a certain number of supporting service facilities such as leisure and recreational facilities, restaurants and cinema and so on. In a modern style, the tower's two pieces of straight glass curtain walls along the river are a metaphor for sails, which form a changeable textural effect with the stagger windows and walls along the residential house side. Vertical line cutting is emphasized in facade design to create a tall and straight building image and make the overall style simple yet grand. The prodium building is decorated with top-grade granite outer walls combined with aluminium alloy grids, which shows not only the modern sense but also the cultural content of the building.

钱江世纪城富丽大厦
Fuli Building of Qianjiang Century CBD

建 设 单 位：杭州富丽达置业有限公司
功 能 用 途：商业、办公
设计/竣工年份：2011年/~
建 设 地 点：浙江省杭州市
总 建 筑 面 积：96 518 m²
总建筑高度/层数：139.9 m/40 F
结 构 形 式：框剪结构
合 作 单 位：北京东方华太建筑设计工程有限责任公司

Owner: Hangzhou Fulida Real Estate Co., Ltd.
Function: Commerce, Office
Design/Completion Year: 2011/~
Construction Site: Hangzhou City, Zhejiang
Total Floor Area: 96,518 m²
Total Height/Floor: 139.9 m/40 F
Structure: Frame-Shear Wall Structure
Cooperation Unit: Beijing Orient Sino-Sun Architectural Design Engineering Co., Ltd.

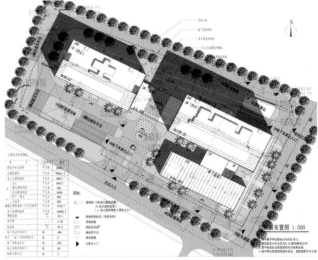

项目位于杭州市萧山区钱江世纪城A-05（1）地块内，地块西侧紧临钱塘江。整个项目由一幢140 m高的超高层、一幢100 m高的高层和一幢40 m高的裙房组成。项目定位为中小企业的商务办公用房。由于基地空间有限，总平面设计不追求复杂的平面布置，这样有利于标准层的空间利用和发挥地下停车库的最大效益。建筑整体造型追求简洁大方，体现超高层建筑的挺拔。建筑外表面材料以铝板和玻璃幕墙为主，竖向的金属线条不但可以为建筑遮阳，而且可以通过光影的变化形成动感的立面线条。

The project is located in Lot A-05(1) of Qianjiang Century CBD in Xiaoshan District, Hangzhou City, next to the Qiantang River in the east. The whole project is composed of a 140 m-high super high-rise building, a 100 m-high high-rise building and a 40 m-high podium building. The project is positioned as a commercial office building for small and medium-sized enterprises. Due to a limited base space, a complex plane layout is not considered in the total plane design, which is good for the space usage of standard floors and because of which underground garage can produce the best possible results. The overall building emphasizes simplicity to show the tallness and straight of the super high-rise building. The outer surface materials of the buildings are mainly aluminium sheets and glass curtain walls. The vertical metal lines can not only shade the sunlight but also form dynamic facade lines with the changes of light and shadow.

杭州江口股份经济合作社等3家经合社商业综合用房
Comprehensive Commercial Buildings of 3 Economic Cooperatives Including Hangzhou Jiangkou Share Economic Cooperative

建设单位：杭州江口股份经济合作社等3家经合社
功能用途：商业、办公
设计/竣工年份：2010年/~
建设地点：浙江省杭州市西湖区
总建筑面积：44 969 m²
总建筑高度/层数：50 m/9 F
结构形式：框架结构

Owner: Hangzhou Jiangkou Economic Cooperative and other 2 Cooperatives
Function: Commerce, Office
Design/Completion Year: 2010/~
Construction Site: Xihu District, Hangzhou City, Zhejiang
Total Floor Area: 44,969 m²
Total Height/Floor: 50 m/ 9 F
Structure: Frame Structure

项目是一座集商场、办公等功能于一体的现代化的综合大厦。项目的建筑主楼的立面采用浅黄色石材、金属百叶、玻璃幕墙与金属构件的有机组合，注重虚实对比，创造了蓬勃向上、富有时代感与活力特征的建筑形象，营造了一个完美的建筑空间。立面重点推敲了构成主楼的各个建筑体块之间的比例和进退、凹凸关系，使整个主楼看起来层次分明，从而展现大气与沉稳的气度。

The project is a modern building complex integrating shopping mall and office. In the selection of the facade style, the dynamic combination of light yellow stone materials, metal louvres, glass curtain walls and metal components is adopted and virtual-real contrast is emphasized, which together create a robust, modern and vigorous building image and a perfect building space. In the facade design, the proportional relation, driving and reversing relation and concave-convex relation between the building masses forming the main building are deliberated, which makes the whole main building look well arranged, grand and steady.

杭州方家畈股份经济合作社商业用房
Commercial Buildings of Hangzhou Fangjiafan Share Economic Cooperative

建设单位：杭州方家畈股份经济合作社
功能用途：商业、办公
设计/竣工年份：2010年/~
建设地点：浙江省杭州市西湖区
总建筑面积：90 696 m²
总建筑高度/层数：50 m/ 9 F
结构形式：框架结构

Owner: Hangzhou Fangjiafan Share Economic Cooperative
Function: Commerce, Office
Design/Completion Year: 2010/~
Construction Site: Xihu District, Hangzhou City, Zhejiang
Total Floor Area: 90,696 m²
Total Height/Floor: 50 m/ 9 F
Structure: Frame Structure

项目位于杭州市转塘镇，定位为转塘镇的区域公共商业和文化中心。设计考虑在有限的土地上，探求能将商业和办公二者结合起来，使得彼此相得益彰而不是互相干扰的最优化的形态组合，努力做到二者既能便利互通又能满足办公区自身相对独立、内向的要求。设计中考虑了全新的视觉效果，立面重点推敲了构成主楼的各个建筑体块之间的比例和进退凹凸关系。在材料的运用上，通过浅黄色石材与玻璃幕墙的组合，形成色彩对比，使得主楼的整体视觉更为灵活丰富。

Located in Zhuantang Town, Hangzhou City, the project is positioned as a regional public commercial and cultural center of Zhuantang Town. The design seeks to combine its commercial district and office district on a limited land area in such a manner that the two can bring out the best in each other but will not interfere with each other, and tries to meet the requirement that the two districts are connected conveniently and the office zone is relatively independent and introverted. In the design, whole new visual effects are considered. In the facade design, the proportional relation, driving and reversing relation and concave-convex relation between the building masses forming the main building are also considered. In the application of materials, the combination of light yellow stone materials and glass curtain wall forms color contrast, to make the overall visual effects of the main building more flexible and richer.

昆明螺蛳湾国际商贸城
Kunming Luoshiwan International Trade City

项目位于昆明市官渡宏仁片区，是一个集现代化、国际化、信息化于一体的综合商贸城。项目总用地面积约为382 hm²，包括总建筑面积约 5 500 000 m²的主体市场（分三期开发）、专业街、金融商务中心、住宅以及医院、学校等各项配套建筑。本案在充分考虑物流、商流、信息流及环境紧密融合的基础上，以"如意结"形的道路结构形成一个优美平稳的规划格局，使高层建筑集群化，利用大型的集合型群落、错落的建筑形态、不同层次空间的串联，营造出一座具有现代商业理念、功能齐全、高效的城中之"城"。

Located in Hongren Plot, Guandu District, Kunming City, the project is a modern, international and information-based comprehensive trade city. The total site area of the project is about 382 hm². The project includes a main market (3 phases) with a floor area of about 5,500,000 m², a specialty street, a financial center, residential houses, hospitals, schools and various supporting buildings. With full consideration of integrating logistics, trade flow and information flow with the environment, the project is provided with a beautiful and stable planning pattern formed by a "traditional Chinese knot" road structure. High-rise buildings are clustered. The large collective clusters, stagger architectural forms and the series connection of different space levels all together create a "city" in the city with modern business ideas, multiple functions and high efficiency.

建设单位：云南中豪置业有限责任公司
功能用途：小商品市场
设计/竣工年份：2009年/2014年
建设地点：云南省昆明市官渡宏仁片区
总建筑面积：8 100 000 m²
总建筑高度/层数：24 m/5 F
结构形式：框剪结构

Owner: Yunnan Zhonghao Real Estate Co., Ltd.
Function: Commodities Market
Design/Completion Year: 2009/2014
Construction Site: Hongren Plot, Guandu District, Kunming City, Yunnan
Total Floor Area: 8,100,000 m²
Total Height/Floor: 24 m/5 F
Structure: Frame-Shear Wall Structure

杭州厚仁商业街
Hangzhou Houren Commercial Street

建设单位：杭州厚仁商业街开发经营有限公司
功能用途：商业、经济型酒店
设计/竣工年份：2009年/2013年
建设地点：浙江省杭州市
总建筑面积：42 951 m²
总建筑高度/层数：36.55 m/10 F
结构形式：框架结构

Owner: Hangzhou Houren Commercial Street Development and Operation Co., Ltd.
Function: Commerce, Economical Hotel
Design/Completion Year: 2009/2013
Construction Site: Hangzhou City, Zhejiang
Total Floor Area: 42,951 m²
Total Height/Floor: 36.55 m/ 10 F
Structure: Frame Structure

　　本项目地处杭州市三墩新城城北居住区，是该居住区的城市功能配套中心，也是三墩新城的第一条大型商业街。办公、经济型酒店和商业设施是本工程的设计主体，营造浓郁的商业氛围并通过合理的交通组织来引导人流是本工程的设计核心。规划采用步行内街的方式来提高整个地块的商业价值和购物的舒适度；办公、经济型酒店采用在底层设集中门厅，在裙房屋顶花园设二门厅的方式，既与商业动线相分离，又改善了办公和居住环境。线形的商业街与点条式的主楼相互穿插，丰富了城市空间。

The project, located in the Chengbei Residential Area in Sandun New Town in Hangzhou City, is the urban function supporting center of the residential area and also the first large commercial street in Sandun New Town. Office, economical hotel and commercial facilities are the design subject of the project, while to create a strong commercial atmosphere and guide stream of people through reasonable traffic organizations is the design core of the project. It is planned to build an inner walking street to increase the commercial value of the whole lot as well as shopping comfort; set a centralized entrance hall on the ground floor in the office and economical hotel, and set a two-door hall on the roof garden of the skirt building, which not only separates the dynamic commercial lines but also improves the office and living environment. The linear commercial street intersects with the point-line-type main building, which enriches the urban space.

宁波江北万达广场
Ningbo Jiangbei Wanda Plaza

建 设 单 位：宁波江北万达置业有限公司
功 能 用 途：商业广场
设计/竣工年份：2009年/2010年
建 设 地 点：浙江省宁波市江北区
总 建 筑 面 积：287 800 m²
总建筑高度/层数：64.7 m/21 F
结 构 形 式：框架结构

Owner: Ningbo Jiangbei Wanda Real Estate Co., Ltd.
Function: Commercial Plaza
Design/Completion Year: 2009/2010
Construction Site: Jiangbei District, Ningbo City, Zhejiang
Total Floor Area: 287,800 m²
Total Height/Floor: 64.7 m/ 21 F
Structure: Frame Structure

项目位于宁波市江北区江北大道以东，云飞路以北，北侧、东侧临宝庆路。本工程规划为城市型商业广场，地上部分建筑面积197 057 m²，地下部分90 757 m²。总图空间布局上分为东、西两大区块，其中西面为大商业区块，地上5层地下2层；东面为商务楼区块，地上21层，地下1层。本案将功能、景观、交通一体化考虑，形成高使用率、高环境品质的集中化城市中心区。

The project is located in the east of Jiangbei Avenue and in the north of Yunfei Road, and adjoins Baoqing Road in the north and Jiangbei District, Ningbo City. The project is positioned as an urban commercial plaza, with a ground floor area of 197,057 m² and an underground area of 90,757 m². In the spatial layout, the project is divided into east and west zones, of which the west zone is the large commercial zone, with 5 floors aboveground and 2 floors underground, and the east zone is commercial building zone, with 21 floors aboveground and 1 floor underground. In the project design, the function, landscape and transportation are considered as a whole, to form a centralized downtown area with high utilization rate and high environmental quality.

文教建筑
Cultural and Educational Building

166	温州大学瓯江学院
168	上海视觉艺术学院国际艺术大师中心
169	余杭区教师进修学校迁建工程
170	杭州师范大学仓前校区二期C区
172	浙江省信息化测绘创新基地
174	慈溪龙山中学迁建工程
175	乐清市实验小学迁建工程
176	慈溪中学
178	阿克苏地区中等职业技术学校
179	乐清市乐成镇第七小学滨海校区
180	乐清市乐成镇第五中学
182	杭州师范大学湘湖校区
184	浙江海洋学院萧山科技学院
186	乐清文化中心
187	遂昌城市文化综合体
188	浙江省地质资料中心
189	嘉兴市图书馆、博物馆二期工程
190	杭州市河道陈列馆
192	宁波市北仑区宁职院图书馆
194	浙江档案馆新馆
195	新疆和田影剧院

166	Oujiang College of Wenzhou University
168	Shanghai Institute of Visual Art International Artists Center
169	Relocation Project of Yuhang Teacher Training School
170	Hangzhou Normal University Cangqian Campus Phase II Zone C
172	Zhejiang Information-based Surveying and Mapping Innovation Base
174	Relocation Project of Cixi Longshan Middle School
175	Relocation Project of Yueqing Experimental Elementary School
176	Cixi Middle School
178	Akesu Secondary Vocational and Technical School
179	Yuecheng No.7 Elementary School Binhai Campus, Yueqing City
180	Yuecheng No.5 Middle School, Yueqing City
182	Hangzhou Normal University Xianghu Campus
184	Zhejiang Ocean University Xiaoshan Campus
186	Yueqing Cultural Center
187	Suichang Urban Cultural Complex
188	Zhejiang Provincial Geoinformation Center
189	Jiaxing Library and Museum Phase II
190	Hangzhou Watercourse Exhibition Hall
192	Beilun Ningbo Polytechnic Library, Ningbo City
194	Zhejiang Provincial Archives New Building
195	Xinjiang Hetian Theater

温州大学瓯江学院
Oujiang College of Wenzhou University

鸟瞰图

教学楼群透视图

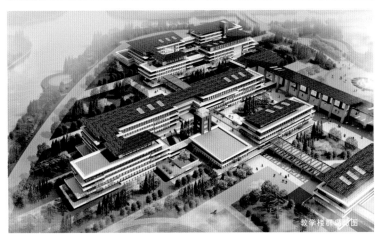

教学楼群鸟瞰图

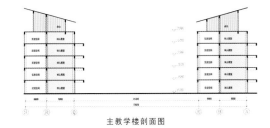

主教学楼剖面图

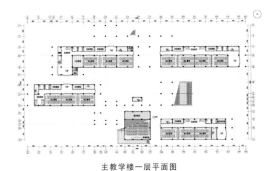

主教学楼一层平面图

主教学楼立面图

本项目的规划设计中，我们以沿望江路布置的中心水系作为校园的生态绿核，结合景观步道，从而形成校园的主体结构。建筑的风格萃取瑞安本地传统建筑精华，通过现代的手法重新构筑，以书院的发展格局作为组团的核心模式，引入"四水归堂"的意境来布置每个组团空间。建筑围绕绿核布置，依据各自功能互为依托，紧密联系，各自形成独立运作的系统，相互间又形成更大的网络空间。造型设计中我们尽量避免整体大体量、大尺度的手法，突出校园建筑的文化底蕴和内涵，为新区构造一个极具江南水乡韵味的大画面。

The planning design of this project takes the central water system along the Wangjiang Road as an ecological green center of the campus. This center and the landscape footpath form the main structure of the campus. The building absorbs local traditional architectural essences of Ruian and is constructed by using modern methods. The layout of traditional school is used as the core plane layout and each block is planned according to the concept of "square central atrium courtyard". Buildings are arranged around the green center and have close functional relation with each other. Each building has its independent operation system and larger network space is formed among them. Efforts are made to avoid large integral volume, so as to highlight the cultural history and connotation of campus building. This creates a large image boasting of architectural style in south of Yangtze River in the Binhai New District.

学生公寓

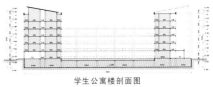
学生公寓楼剖面图

学生公寓楼南立面图

会堂

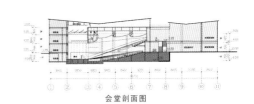
会堂剖面图

会堂立面图

学院楼

图文信息中心

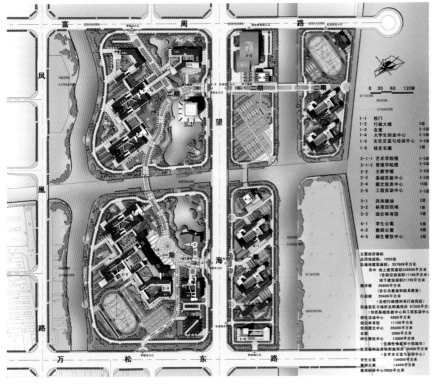

建设单位：温州侨大发展有限责任公司
功能用途：教学
设计/竣工年份：2013年/～
建设地点：浙江省瑞安市
总建筑面积：357 886 m²
总建筑层数：10 F
结构形式：框架结构

Owner: Wenzhou Qiaoda Development Co., Ltd.
Function: Education
Design/Completion Year: 2013/~
Construction Site: Ruian City, Zhejiang
Total Floor Area: 357,886 m²
Total Floor: 10 F
Structure: Frame Structure

上海视觉艺术学院国际艺术大师中心
Shanghai Institute of Visual Art International Artists Center

建设单位：上海视觉艺术学院
功能用途：教学
设计/竣工年份：2010年/2012年
建设地点：上海市松江区
总建筑面积：24 349 m²
总建筑高度/层数：68.2 m /15 F
结构形式：框剪结构

Owner: Shanghai Institute of Visual Art
Function: Education
Design/Completion Year: 2010/2012
Construction Site: Songjiang District, Shanghai
Total Floor Area: 24,349 m²
Total Height/Floor: 68.2 m /15 F
Structure: Frame-Shear Wall Structure

本项目又名国际大师中心，主要功能是为艺术大师们提供创意工作与交流学习的场所。建筑的造型采用前低后高的体块处理方式，表现了简洁优雅的建筑风格，立面造型立意取自于钢琴。建筑用优雅的白色墙体线条勾勒出挺拔、典雅的建筑形象，建筑南北立面错落的深色窗格仿佛黑亮的钢琴键上跳动着的音符。南入口的雨篷及入口造型犹如一片撑开的钢琴盖板和支架。建筑的北立面及侧面设计成浮雕的形式，犹如信息时代的数码线条，又仿佛音乐般跳跃。整座建筑色彩对比强烈、格调明快，运用钢琴、数码的寓意突出艺术的主题。

This project, also called International Masters Center, provides creative working, communication and learning spaces for artistic masters. The building is composed of lower front volume and higher back volume to realize simple but elegant building shape. The facade is designed on the "piano" concept. Elegant white wall profile highlights a tall, straight and classic building image. Stagger dark muntin on south and north facades look like bright black piano keys. The awning over south entrance and the entrance itself are just like an unfolded piano cover plate and supporting element. North elevation and side facade of the building are designed into embossment, looking like digital lines in modern information age and bumping music notes. The whole building boasts of intensive color contrast and vigorous appearance, and highlights the artistic theme through piano and digital elements.

余杭区教师进修学校迁建工程
Relocation Project of Yuhang Teacher Training School

建设单位：余杭区教师进修学校
功能用途：教学
设计/竣工年份：2013年/~
建设地点：浙江省杭州市余杭区
总建筑面积：26 429 m²
总建筑高度/层数：62.7 m/15 F
结构形式：框架结构

Owner: Yuhang Teacher Training School
Function: Education
Design/Completion Year: 2013/~
Construction Site: Yuhang District, Hangzhou City, Zhejiang
Total Floor Area: 26,429 m²
Total Height/Floor: 62.7 m/15 F
Structure: Frame Structure

项目位于杭州市余杭区临平镇红丰立交南侧，校园划分为教学行政楼、网球中心两个部分。其中教学区位于西侧，网球中心位于东侧，两者间又以通廊相联系，使之自然地融合成一个整体。
教学主楼及其裙房部分采用了大气的方格网式立面和冷暖结合的色调，赋予建筑沉稳而不沉闷、明亮却不透亮的整体感。网球中心的建筑肌理采用一种网状折形结构，呈现积极向上的态势，在方正中不失张扬。

The project is located at the south of Hongfeng Overpass in Linping Town of Yuhang District, Hangzhou City, and the campus is divided into teaching and administrative building and tennis center. The teaching zone is planned on the west part and tennis center is on the east part. They are connected through a corridor into an integral body.
The main structure and podium structure of the teaching building adopt generous square lattice facade and mixture of cold and warm colors, to create a solemn but vigorous building image and bright but opaque volume. The tennis center is a folded net structure, embodying positive and booming, regular and fashionable image.

杭州师范大学仓前校区二期C区
Hangzhou Normal University Cangqian Campus Phase II Zone C

通过对"细胞"形态、结构、生长、繁殖、遗传、变异、相聚、相离等规律性的提炼,将其作为建筑语言应用到建筑群体的布局与组织中。不同体量的功能院落体块,依循各自功能分区和相互不同的关联度,因地就势、起承转合,形成既能延续拓展"湿地书院"的主文脉,又独具"医药与生命科学学部"特色的建筑形态——"生命聚落"。在主体建筑造型上,我们效法"叠石为山",建筑立面采用横向挑台,强调了建筑横向线条的大气舒展,顶部采用退台、绿坡等手法,撷取山形意向;起伏飘逸的绿色网膜体系则宛若彩云,与基地内丰富的水网、河道、树木资源,共同勾勒出一幅"山、水、云、林"的人文美景,与一期的"水院"主题相映成趣。

After analysis on shape, structure, growth, reproduction, heredity, variation, aggregation and separation of "cell", the cell concept is used as an architectural language for layout and organization of building cluster. Functional courtyards of different sizes are planned according to their functions, relations and geological conditions. This results a "life colony" building prototype which widens the main cultural clue of "wetland campus" and has unique features of "medical and life science division". The building shape imitates "stone-stacked mountain" and transverse cantilever plate on facade highlights the generous and unfolded transverse profile of the building. Terraced and slope roof looks like waving mountain, and floating green screen system is just like floating colorful clouds; the building cooperates with abundant waterscapes, watercourse and trees on the site to draw a beautiful humanism picture of "mountain, water, cloud and forest", reflecting the "water courtyard" theme of Phase I.

建 设 单 位：杭州师范大学
功 能 用 途：教育
设计/竣工年份：2012年/～
建 设 地 点：浙江省杭州市仓前镇
总建筑面积：242000 m²
总建筑高度/层数：60 m / 11 F
结 构 形 式：框架结构

Owner: Hangzhou Normal University
Function: Education
Design/Completion Year: 2012/~
Construction Site: Cangqian Town, Hangzhou City, Zhejiang
Total Floor Area: 242,000 m²
Total Height/Floor: 60 m / 11 F
Structure: Frame Structure

浙江省信息化测绘创新基地
Zhejiang Information-based Surveying and Mapping Innovation Base

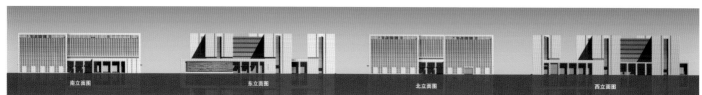

项目兼具科研、行政与文化功能。科研型建筑造型动感，反映科技精神；行政型建筑造型稳重而严肃；文化建筑如博物馆，形态具有学术性与纪念性特质。"天之度、地之尺"是本案的设计理念，既在抽象层面体现测绘工作的严谨性和重复性，又在具象层面表现建筑设计的艺术性与创新性。布局上我们采用通风采光较好的板式建筑，通过单向性横条式的重复、错动、长短变化等，形成一种理性却不失变化的布局形态。面向南侧主入口广场的博物馆设置主题雕刻，强化了测绘信息博物馆的功能与特点，同时为以理性和刚性为主的整体建筑群加入活跃的元素。

The project provides various functions such as scientific research, administration and culture, etc. Scientific research building uses dynamic shape to reflect the scientific and technological spirits; administration building has solemn and elegant appearance; cultural buildings such as museum are often designed with academic and commemorative features. It is based on the design concept of "boundless skyline and natural terrain", namely expressing the preciseness and reproducibility of surveying and mapping work on the abstract aspect, and indicating the artistic and innovative features of architectural design on the concrete aspect. Slab-type building layout is adopted to guarantee favorable ventilation and lighting effects; an appropriate and diversified layout is realized through repetition, staggering and different lengths of unidirectional transverse strips. Themed sculpture is designed for the museum facing the main entrance square at south, to highlight functions and characteristics of surveying and mapping information museum and to infuse vigorous elements into the integral building cluster which boasts of reasonable and rigid appearance.

建设单位：浙江省测绘与地理信息局
功能用途：科教基地
设计/竣工年份：2012年/～
建设地点：浙江省杭州市余杭区
总建筑面积：73 460 m²
总建筑高度/层数：34 m /7 F
结构形式：框剪结构

Owner: Zhejiang Adminstration of Surveying, Mapping and Geoinformation
Function: Scientific Teaching Base
Design/Completion Year: 2012/~
Construction Site: Yuhang District, Hangzhou City, Zhejiang
Total Floor Area: 73,460m²
Total Height/Floor: 34 m /7 F
Structure: Frame-Shear Wall Structure

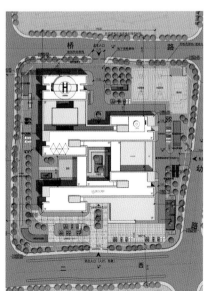

慈溪龙山中学迁建工程
Relocation Project of Cixi Longshan Middle School

建 设 单 位：慈溪市教育发展有限公司
功 能 用 途：教育
设计/竣工年份：2012年/~
建 设 地 点：浙江省慈溪市龙山镇
总 建 筑 面 积：42 000 m²
总建筑高度/层数：27.5 m /7 F
结 构 形 式：框架结构

Owner: Cixi Educational Development Co., Ltd.
Function: Education
Design/Completion Year: 2012/~
Construction Site: Longshan Town, Cixi City, Zhejiang
Total Floor Area: 42,000 m²
Total Height/Floor: 27.5 m/7 F
Structure: Frame Structure

本项目位于慈溪市龙山镇龙山新城，总用地面积约80 000 m²。

具有历史感的校园建筑的"美"，并不在于新颖、独特的外在造型，更在于其折射出历经沧桑后逐渐完善的文化内涵。在整个规划中，我们以传统的空间形态将各个建筑组织起来，再通过庭院尺度的大小来进行主次空间的区分；通过架空层、步行道、连廊等交通空间，给师生们展开一幅优美的画卷。相互环绕、穿插的大小庭院，加上变化的交通空间，形成校园内既有近人尺度，又富有层次的空间形态。

The project is located in Longshan New Town of Longshan Town, Cixi City, covering a total site area of about 80,000 m².
The "esthetics" of historical campus building is determined not only by creative and unique external appearance, but also more on its historical cultural connotation. In this project, buildings are organized through traditional space layout and functional importance is divided according to different courtyard sizes; lifted path, footpath and corridor unroll a beautiful picture for students and teachers. Crossed courtyards of different sizes and diversified communication spaces form human scale and sophisticated space layout.

乐清市实验小学迁建工程
Relocation Project of Yueqing Experimental Elementary School

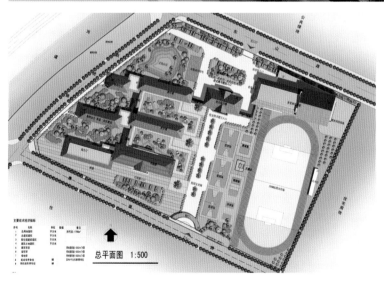

总平面图 1:500

建设单位：乐清市实验小学
功能用途：教育
设计/竣工年份：2011年/2013年
建设地点：浙江省乐清市
总建筑面积：23 862 m²
总建筑高度/层数：18.9 m/5 F
结构形式：框架结构

Owner: Yueqing Experimental Elementary School
Function: Education
Design/Completion Year: 2011/2013
Construction Site: Yueqing City, Zhejiang
Total Floor Area: 23,862 m²
Total Height/Floor: 18.9 m/5 F
Structure: Frame Structure

学校共设置36间标准尺寸的教室，实验教室、办公区、图书馆、食堂、游泳馆、篮球馆等各个功能空间有分有合，既联系便捷，又动静分明，互不干扰。

弧形阵列的校园大门是整个校区序列的起点，形成建筑空间上第一个高潮。经过校园文化轴，学校空间在此进一步延伸，向人们展示新校区独有的建筑空间。行政综合楼在空间上是承上启下的枢纽，立面形式上也是校区整体风格传承的重点。底层共享空间连接了整个校区的绝大多数功能体，是新校区使用功能与空间景观的双重中心。食堂和综合楼围合而成的后广场，成为整个序列的结束。

This school has various functional spaces, such as 36 standard classrooms, experimental classrooms, offices, library, dining hall, swimming pool and basketball hall, etc., which are connected with or separated from each other, providing convenient and undisturbed communication.
The arched matrix gate is the starting point of the campus, forming the first climax of building space. Passing through the campus cultural axis, the campus space is further developed to show the unique building space of the new campus. The administration general building plays a space hub role on the campus and facade is also an important element inheriting the overall historical style of the new campus. The public space on ground level connects most of functional areas of the campus, it is both functional center and landscape center. The back square enclosed by dining hall and general building is the end point of the campus.

慈溪中学
Cixi Middle School

主校门

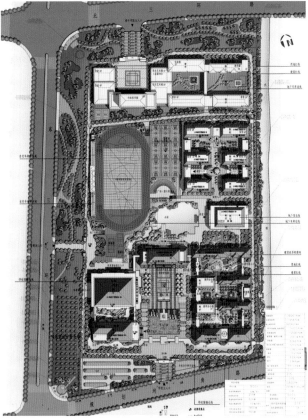

行政图书楼

内庭院

项目位于慈溪市白沙路，总用地面积约138 736 m²，拟建60个班的寄宿制高级中学。

在整个规划设计中，通过庭院尺度的大小来进行主次空间的区分。架空步道、连廊等交通空间的运用，环绕、穿插的大小庭院，形成校园内既有近人尺度，又富有层次的空间形态。"亦园亦院"的布局手法运用于慈溪中学新校园的排布中。

新校区建筑群沿袭历史文脉的印记。各个功能区块都以风雨连廊相连，连廊入院，连廊入楼，既方便了学生、老师便捷地出入各个楼体之间，又增加了校园内园林景致的趣味性。

The project is located on Baisha Road in Cixi City, covering a total site area of about 138,736 m². It will be a boarding senior high school having 60 classrooms.

The overall planning design differentiates the functional importance according to courtyard size. Lifted footpath, corridor and crossed courtyards of different sizes create diversified spaces in the campus. "garden and courtyard" layout is used in this new campus.

The building cluster in this new campus will inherit the historical culture. Various functional areas are connected with each other through roofed corridors which penetrate into courtyards and buildings. This layout not only provides convenience for students and teachers to enter or exit the buildings, but also add more interest into the gardening landscape on the campus.

建设单位：慈溪市教育发展有限公司
功能用途：教育
设计/竣工年份：2012年/～
建设地点：浙江慈溪市
总建筑面积：118 000 m²
总建筑高度/层数：38.5 m /9 F
结构形式：框架结构

Owner: Cixi Educational Development Co., Ltd.
Function: Education
Design/Completion Year: 2012/~
Construction Site: Cixi City, Zhejiang
Total Floor Area: 118,000 m²
Total Height/Floor: 38.5 m /9 F
Structure: Frame Structure

体育馆

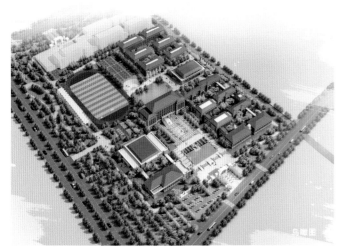
鸟瞰图

艺术楼

教学楼连廊

宿舍及下沉广场

教学楼

阿克苏地区中等职业技术学校
Akesu Secondary Vocational and Technical School

建设单位：阿克苏地区中等职业技术学校
功能用途：教育
设计/竣工年份：2011年/~
建设地点：新疆阿克苏市
总建筑面积：114 000 m²
总建筑高度/层数：21 m /5 F
结构形式：框架结构

Owner: Akesu Secondary Vocational and Technical School
Function: Education
Design/Completion Year: 2011/~
Construction Site: Akesu City, Xinjiang
Total Floor Area: 114,000 m²
Total Height/Floor: 21 m /5 F
Structure: Frame Structure

项目位于阿克苏地区纺织工业园区的地区实验林场东南新规划区内，用地面积606 206 m²。学校预期在校学生8 000人，总投资3亿元人民币，由一幢教学楼、三幢实训楼、一幢图书信息楼、八幢学生宿舍、两个学生食堂、一幢生活服务楼和配套体育运动场地构成。

校区采用开放式网络状、绿化生态系统。南北向的弧形通道是校区的主干道路，将校区划分为教学区和生活区两大部分，动静分明。校区内形成一纵一横两道景观轴。纵向景观轴为沿河绿化带，贯穿校前区广场。横向景观轴起点为图书信息楼南面的景观亭，终点为宿舍区组团的中心机房。此轴线连通了教学区和生活区，使两区块形成一个有机的整体。

The project is located in the southeast new planning zone of regional experimental forest in the textile industrial park of Akesu City, and covers a site area of 606,206 m². The school is planned to accept 8,000 students and its total investment is RMB 300 millions. The school is composed of one teaching building, three training buildings, one library and information building, eight students' dormitory buildings, two canteens, one living service building and supports field, etc.
The campus adopts open network green ecological system. The arched path stretching from north to south is the main road in the campus and divides the campus into teaching zone and living zone. There is one transverse axis and one longitudinal landscape axis in the campus. Waterfront green belt, passing through the front square, forms the longitudinal landscape axis; the transverse landscape axis starts from the landscape booth at south of the library and information building and ends by the central machine room in the dormitory zone. It connects the teaching zone and the living zone into an organic whole.

乐清市乐成镇第七小学滨海校区
Yuecheng No.7 Elementary School Binhai Campus, Yueqing City

建 设 单 位：乐清市乐成镇第七小学
功 能 用 途：教育
设计/竣工年份：2011年/～
建 设 地 点：浙江省乐清市
总 建 筑 面 积：19 100 m²
总建筑高度/层数：21 m/5 F
结 构 形 式：框架结构

Owner: Yuecheng No.7 Elementary School, Yueqing City
Function: Education
Design/Completion Year: 2011/~
Construction Site: Yueqing City, Zhejiang
Total Floor Area: 19,100 m²
Total Height/Floor: 21 m/5 F
Structure: Frame Structure

本项目位于乐清市中心区II-a6地块，规划总用地面积31 349 m²，拟建36个教学班、可容纳在校生1620人的一类小学。

校园主要分为教学区、生活运动区和行政办公区，三者通过绿化广场有机地围合成一个校园空间。在造型设计上，行政办公区主要形成门的形式，寓意知识之门。教学区则通过坡屋顶的元素、竖向线条、彩色铝板、木遮阳百叶穿插，丰富了立面层次。在主立面的处理上，以橘红色面砖为主调，加以玻璃的对比。整个屋面犹如一条飞扬的红领巾，屋面板设计上强调建筑"飘"的含义，体现轻盈舒展的建筑形体，同时又起到了通风隔热的效果。

The project is located on plot II-a6 in the central district of Yueqing City, covering a total site area of 31,349 m². A grade-I elementary school is planned to be built here, including 36 classrooms for 1,620 students.

The campus is mainly divided into teaching zone, living and sports zone and administrative office zone, which are connected through a green square into an organic campus. In the administrative office zone, the building is planned into a door shape, embodying the door of knowledge. In the teaching zone, the building adopts inclined roof, vertical line, colorful aluminium panel and wooden sunshade shutter to form sophisticated facade image. The main facade is decorated with jacinth tile and glass. The whole roof looks like a piece of flying red scarf. It focuses on the "flying" image, to embody light and unfolded building shape and to realize favorable ventilation and insulation effects.

乐清市乐成镇第五中学
Yuecheng No.5 Middle School, Yueqing City

项目位于乐清市乐成镇,学校的规划设计遵循三大特色:人文校园、生态校园、数字校园。

在校区建设中充分体现出地域文化的内涵,通过对乐清市牌楼文化的挖掘,将其融入校园建设中,进一步传承和发扬。设计体现了校园建筑与周围环境的和谐统一,校园周围背山面水,内有庭院相连,山、水、园三趣相合,内外相呼应,既环境舒适又节能环保,尽显可持续发展的生态特色。同时为了满足现代教育需求,通过数字手段,使校园内的管理、教育、教学手段更现代化、数字化和智能化。

The project is located in Yuecheng Town, Yueqing City, and it is planned and designed on the following three main features, including humanism campus, ecological campus and digital campus.
Local cultural connotation is fully expressed during construction of this campus. For example, local archway culture of Yueqing City is studied and introduced into the campus, so as to further inherit and promote the architectural culture. The design embodies harmonious and uniform relation between campus building and contextual environment. The campus enjoys beautiful mountainous landscape and waterscape elements. On the campus, courtyards are connected with each other, mountains, waterscape and gardening landscape are integrated to realize dialogue between internal and external spaces. This environment-friendly and energy-saving design produces a sustainable ecological campus. Digital means are adopted to satisfy modern educational requirements and to realize more modern, digital and intelligent management, education and teaching on campus.

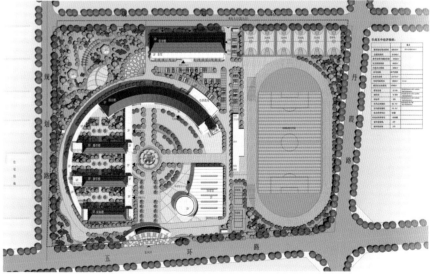

建设单位：乐清市乐成镇第五中学	Owner: Yuecheng No.5 Middle School, Yueqing City

建设单位：乐清市乐成镇第五中学
功能用途：教育
设计/竣工年份：2011年/2013年
建设地点：浙江省乐清市乐成镇
总建筑面积：31 540 m²
总建筑高度/层数：22.2 m /6 F
结构形式：框架结构

Owner: Yuecheng No.5 Middle School, Yueqing City
Function: Education
Design/Completion Year: 2011/2013
Construction Site: Yuecheng Town, Yueqing City, Zhejiang
Total Floor Area: 31,540 m²
Total Height/Floor: 22.2 m / 6 F
Structure: Frame Structure

杭州师范大学湘湖校区
Hangzhou Normal University Xianghu Campus

本项目位于杭州市萧山区，占地面积约265000 m²。整个校区布局为"一心、两轴、四区"的规划结构。"一心"：由最具活力的图书馆、培训行政综合楼组成中心区。"两轴"：西北生活区与东南教学区，穿越中心开放区，形成符合日常行为的生活轴；南面经由校前广场、中心景观和图书馆与培训行政综合楼限定的中心广场形成300 m长的景观主轴。"四区"：两条轴线将校园合理地划分为四个区，即教学区、生活区、运动区和后勤区。

"百年荟萃，杭派印象"，设计秉承杭州师范大学百年历史文脉及人文精神，创造一系列展现杭城魅力以及校园历史文脉与校园精神的环境场所，实现杭州师范大学传统人文精神的再现，唤起师生的共鸣。

This project is located in Xiaoshan District, Hangzhou City, covering a site area of about 265,000 m². The whole campus follows a planning layout of "one center, two axes and four zones". "One center" refers to a central zone composed of the most vigorous library and training administration general building. "Two axes" refer to a living axis formed by the northwest living zone and southeast teaching zone passing through the central open area, and a 300 m long landscape axis at south formed by front square, central landscape and central square enclosed by library and training administration general building. "Four zones" are reasonably divided by the two axes, including teaching zone, living zone, sports zone and logistic zone.

Based on "centennial essence and Hangzhou impression"— the centennial cultural history and humanism spirit of Hangzhou Normal University, this design successfully creates a series of environmental locations showing the charm of Hangzhou City, cultural history and campus spirit, highlights the traditional humanism spirit of Hangzhou Normal University, and evokes teachers' and students' deep memory.

建 设 单 位：杭州师范大学
功 能 用 途：教育
设计/竣工年份：2011年/～
建 设 地 点：浙江省杭州市萧山区
总建筑面积：240 000 m²
总建筑高度/层数：49 m /11F
结 构 形 式：框架结构

Owner: Hangzhou Normal University
Function: Education
Design/Completion Year: 2011/~
Construction Site: Xiaoshan District, Hangzhou City, Zhejiang
Total Floor Area: 240,000 m²
Total Height/Floor: 49 m/11 F
Structure: Frame Structure

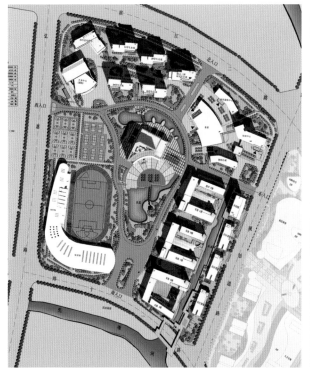

浙江海洋学院萧山科技学院
Zhejiang Ocean University Xiaoshan Campus

本项目位于杭州市萧山高教园区东北侧。设计以基地环境为出发点，以西侧为学校主入口，并将其扩大为校前广场。利用原有地形中的河道将地块划分为东西两区。西侧地块自南向北依次布置有体育运动区、教学实训楼、图书行政楼、实训楼、实训基地、教学科研创新基地、养殖基地、萧山科学技术研究院；后勤生活保障区设置于东区，自南向北依次为食堂、学生宿舍、研究生宿舍、教师宿舍与招待所。校园以中心景观大道结合丰富的水体为主线、院落体系为辅线来组织空间，形成了大气的中心景观通廊和精致的院落组合。设计展现了空间层次丰富、景观环境幽雅、交流氛围浓郁、整体协调秀美的校园环境。

This project is located at northeast part of Advanced Educational Zone in Xiaoshan District, Hangzhou City. Based on contextual environment, this scheme plans the main entrance at west, which is expanded into a front square. The site is divided by the existing watercourse into east and west zones. The west zone includes the following functions distributed from south to north: sports area, teaching and training building, library and administrative building, training building, training base, teaching, scientific research and innovation base, cultivation base, Xiaoshan Institute of Scientific and Technical Research; logistic and living zone is planned in the east zone, including the following functions from south to north: dining hall, students' dormitories, graduate students' dormitories, teacher's dormitories and hotel. The campus space is organized according to a main clue of central landscape boulevard and abundant water systems and an auxiliary clue of courtyard layout, creating a generous central landscape corridor and delicate combination of courtyards. The design scheme boasts of diversified layouts, comfortable and elegant landscape environment, intensive communication atmosphere and harmonious campus environment, etc.

建设单位：浙江海洋学院
功能用途：教育
设计/竣工年份：2010年/2013年
建设地点：浙江省杭州市萧山高教园区
总建筑面积：195 777 m²
总建筑高度/层数：24 m/6 F
结构形式：框架结构

Owner: Zhejiang Ocean University
Function: Education
Design/Completion Year: 2010/2013
Construction Site: Advanced Educational Zone, Xiaoshan District, Hangzhou City, Zhejiang
Total Floor Area: 195,777 m²
Total Height/Floor: 24 m/6 F
Structure: Frame Structure

乐清文化中心
Yueqing Cultural Center

A-d4地块：图书馆、博物馆项目
A-d7地块：文化公园项目
A-d11地块：文化综合体项目
整体总平面

建 设 单 位：乐清中心区开发建设委员会、乐清文广新局
功 能 用 途：文化综合体
设计/竣工年份：2012年/～
建 设 地 点：浙江省乐清市
总 建 筑 面 积：101 300 m²
总建筑高度/层数：20.5 m /5 F
结 构 形 式：框架结构

Owner: Yueqing Central District Development and Construction Committee/ Yueqing Cultural, Broadcasting and News Bureau
Function: Cultural Complex
Design/Completion Year: 2012/~
Construction Site: Yueqing City, Zhejiang
Total Floor Area: 101,300 m²
Total Height/Floor: 20.5 m /5 F
Structure: Frame Structure

本项目包含A-d4、A-d7和A-d11三个地块。A-d4地块为图书馆、博物馆项目。图书馆总藏书量135万册，总阅览座位1 200个；博物馆为大型馆。A-d7地块为文化公园，规划用地面积53 708 m²。A-d11地块为文化综合体项目，包括文化馆、1000座剧院、600座音乐厅、影院（包括两个200座、三个100座、三个50座的放映厅）以及群艺馆等。

图书馆和博物馆建筑造型犹如叠石形成的山体，位于文化公园西侧。文化综合体建筑造型犹如经流水打磨的玉石镶嵌于原石之中，位于文化公园东侧。三个地块相互呼应，融为一体，体现"山水乐清"的设计理念。

This project is located on three plots, including A-d4, A-d7 and A-d11. Plot A-d4 includes library and museum functions. The library has 1.35 million books and provides 1,200 reading seats; the museum has large building volume. Plot A-d7 is planned as a cultural park, covering a planning site area of 53,708 m². Plot A-d11 is planned as a cultural complex which is composed of cultural hall, 1,000-seat theater, 600-seat concert hall, cinema (including two 200-seat hall, three 100-seat hall and three 50-seat hall) and collective arts hall, etc.
The library and museum, located at west of the cultural park, are just like a stone-stacked mountain. The cultural complex building is located at east of the cultural park and looks like a water-polished jade embedded in a piece of natural stone. These three plots respond to each other and are integrated into a whole, embodying the design concept of "mountainous and waterfront Yueqing City".

遂昌城市文化综合体
Suichang Urban Cultural Complex

建设单位：遂昌县城市建设发展有限公司
功能用途：文化综合体
设计/竣工年份：2013年/~
建设地点：浙江省丽水市遂昌县
总建筑面积：41 320 m²
总建筑高度/层数：21 m/5 F
结构形式：框架结构

Owner: Suichang Urban Construction and Development Co., Ltd.
Function: Cultural Complex
Design/Completion Year: 2013/~
Construction Site: Suichang County, Lishui City, Zhejiang
Total Floor Area: 41,320 m²
Total Height/Floor: 21 m/5 F
Structure: Frame Structure

本项目位于遂昌县新城核心区位，总用地面积为22 738.5 m²，是一座集文化馆、青少年宫、工人文化宫、图书馆、档案馆、城市展览馆及遂昌剧院等功能于一体的文化综合体。

设计从"天圆地方"的传统文化精髓出发，整座建筑以呈现"方"形的完整建筑形态为主导。以黑白灰经典搭配构成的建筑立面，和谐统一，使整个立面富有艺术性与观赏性。中心的大中庭与建筑体量相呼应，在不同时段阳光的照射下，光影就像跳动的音符，富有节奏感。

经过推敲后的建筑，不仅形象舒展、尺度合理，而且与周边建筑相协调，营造出丰富的文化空间和建筑氛围。

This project is located in the central district of new town of Suichang County, covering a total site area of 22,738.5 m². It is a cultural complex integrating cultural hall, youth palace, workers' cultural palace, library, archives, urban exhibition hall and Suichang Theater, etc.

The design scheme follows the traditional cultural essence of "round sky and square earth" and focuses on "square" building volume. Building facades are designed in classic combination of black, white and grey colors, to create a harmonious, uniform, artistic and esthetic building appearance. The large central atrium responds to the solid building volume, producing jumping music notes of light and shadow at different periods of a day.

After careful analysis, this design realizes elegant building image and reasonable building volume, boasting harmonious relation with surrounding buildings and highlighting diversified cultural atmosphere and building spaces.

浙江省地质资料中心
Zhejiang Provincial Geoinformation Center

建设单位：浙江省地质档案馆
功能用途：档案馆、博物馆
设计/竣工年份：2012年/～
建设地点：浙江省杭州市萧山区
总建筑面积：34 340 m²
总建筑高度/层数：71.4 m/15 F
结构形式：框筒结构

Owner: Zhejiang Provincial Geological Archives
Function: Archives, Museum
Design/Completion Year: 2012/～
Construction Site: Xiaoshan District, Hangzhou City, Zhejiang
Total Floor Area: 34,340 m²
Total Height/Floor: 71.4 m/15 F
Structure: Frame-Tube Structure

　　本项目位于杭州市萧山区金山路东侧，总用地面积为16 033 m²。建成后是省内地质资料最齐全的专业地质资料档案馆。
　　设计以地质现象中最有代表性的"断层"和"节理"为设计元素，提取其中精华，通过建筑体块的分解错置以及大块石材的运用，体现"石之记忆"这一理念。根据地块狭长的特征，我们将主楼布置在基地最北侧，舒展的裙房结合稳重的底座肌理，与南侧端部的斜切元素，构筑成一幅正乘风破浪前行的"地质航母"形象。

The project is located on the east side of Jinshan Road in Xiaoshan District, Hangzhou City, covering a site area of 16,033m². After completion, it will be the most professional geoinformation archives possessing the most geological data in Zhejiang Province.
The design is based on "faultage" and "joint", which are the most representative geological phenomenon, to extract the essential elements and to embody the concept of "stone memory" through stagger position of building volumes and large stone. The site is long and narrow, so the main building is planned at the north end of the site. Unfolded podium building standing on stable base cooperates with the chamfered elements at south end to create an image of sailing "geological aircraft carrier".

嘉兴市图书馆、博物馆二期工程
Jiaxing Library and Museum Phase II

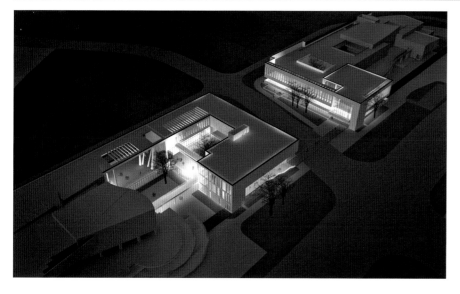

建设单位：嘉兴市图书馆、博物馆
功能用途：图书馆、博物馆
设计/竣工年份：2011年/～
建设地点：浙江省嘉兴市
总建筑面积：17 205m²
总建筑高度/层数：18.9m/4F
结构形式：框架结构

Owner: Jiaxing Municipal Library and Museum
Function: Library, Museum
Design/Completion Year: 2011/~
Construction Site: Jiaxing City, Zhejiang
Total Floor Area: 17,205 m²
Total Height/Floor: 18.9 m/4 F
Structure: Frame Structure

　　项目位于嘉兴市中心，东至海盐塘路，南临图书馆一期，西至七一广场。在本工程的扩建设计和立面改造设计中，我们采用整体规整而局部灵活的布局方式，旨在创造该区域的文化肌理，建立该区域建筑与自然环境的和谐统一。
　　在博物馆和图书馆立面设计中，我们将新建筑立面延伸、线、面、体简单而丰富的组合使原有的建筑立面重新焕发出新的生机，在完成老建筑立面改造的同时也很好地完成了新旧建筑的对话。

The project is located at the central district of Jiaxing City and is adjacent to Haiyantang Road at east, Library Phase I at south and Qiyi Square at west. Extension design and facade reconstruction design of this project follow the concept of integral regularity and local flexibility, to create cultural texture of this district and to realize harmonious combination between buildings and natural environment.
In the facade design of museum and library, facades of the new building are extended. Simple and diversified combination of line, surface and volume infuses vigorous life into the building facades, not only successfully rebuilding the old building facades, but also realizing dialogue between the new and old buildings.

杭州市河道陈列馆
Hangzhou Watercourse Exhibition Hall

项目位于杭州市拱墅区上塘河与姚家坝交叉口西南侧，原名杭州市城区河道"水文化"文明基地，浓缩了杭州因水而生的灵性和风采，宣传水环境整治，弘扬水文化。项目总用地面积9 240 m²，地形呈长方形。

设计充分考虑古老运河的文化底蕴和江南传统的庭院建筑形式。将景观从墙外引入建筑内，使陈列馆的三个体块和配套用房形成一个围合空间，运河绿化带和陈列馆的小庭院用通透的玻璃体联系，内外景观互相渗透穿插，营造出共享的绿化空间。整个项目采取局部集中的建筑布局，陈列馆的形式采用类似官府的城墙元素，建筑的风格蕴含中式元素，运用现代的工艺技术。设计营造了一个环境舒适、生态复合、环保科技的青少年教育基地。

This project is located at southwest of the intersection between Shangtang River and Yaojia Dam in Gongshu District, Hangzhou City. Formerly called urban watercourse "water culture" civilization base, it extracts the urban spirit and waterscape of Hangzhou, and propagandizes water environment treatment and water culture. The rectangular site covers a total area of 9,240 m². The design gives full consideration to the ancient canal culture and traditional courtyard prototype in regions south of the Yangtze River. Outside landscape is introduced into the building; three volumes of the exhibition hall and auxiliary facilities enclose a central space; canal green belt and small courtyard of the exhibition hall are connected by glass system to realize interpenetration of inside and outside landscape and to create a shared green space. The whole project adopts a locally concentrated building layout and the exhibition hall uses rampart elements of feudal official building, boasting of Chinese-style architectural elements and modern techniques. This design successfully creates a youth education base highlighting comfortable, ecological, environment-friendly, scientific and technological features.

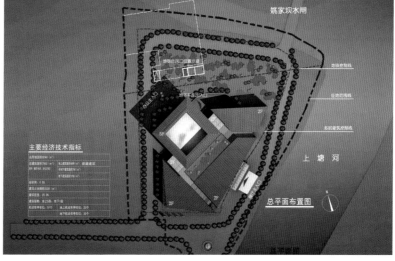

建设单位：杭州市市区河道建设中心	Owner: Hangzhou Urban Watercourse Construction Center
功能用途：陈列馆	Function: Exhibition Hall
设计/竣工年份：2010年/～	Design/Completion Year: 2010/～
建设地点：浙江省杭州市拱墅区	Construction Site: Gongshu District, Hangzhou City, Zhejiang
总建筑面积：7 563 m²	Total Floor Area: 7,563 m²
总建筑高度/层数：20 m/5 F	Total Height/Floor: 20 m/5 F
结构形式：框架结构	Structure: Frame Structure

宁波市北仑区宁职院图书馆
Beilun Ningbo Polytechnic Library, Ningbo City

项目位于北仑区的西南侧,由北仑区公共图书馆与宁波职业技术学院高校图书馆两部分组成。

学校图书馆的人行主入口设在学校行政楼的东北侧,通过扶桥可直接到达图书馆的二层。公共图书馆设在基地北部共2层,通过架空层及灰空间的过渡,与设在基地南侧的学校图书馆连为一个整体。建筑风格着力在整体和谐的基础上,分别突出学校图书馆及公共图书馆各自的特点,强调两馆的可识别性。建筑4~5层以陶土板和玻璃幕墙混合立面为主体。图书馆南外立面设一系列构件,由铝合金材料制成的充满古典韵味的通透格窗,错落灵动,间以少量的古文字,营造出独特的文化氛围。

The project is located at southwest area of Beilun District and is composed of Beilun Public Library and Ningbo Polytechnic Library.

The main entrance of the campus library is planned at northeast of the administration building and people could directly reach 2F through a pedestrian bridge. The 2-storey public library is planned at north part of the site and it is connected with the campus library through transition of overhead layer and grey space. Overall harmonious appearances of these two libraries highlight their own features and identifiability. 4F-5F facades are decorated with ceramic panel and glass curtain wall. A series of components is designed on the south facade of library, aluminium lattice windows are staggered regularly and are dotted with a few ancient characters, creating an unique cultural atmosphere.

建设单位：宁波市北仑区建筑工务局
功能用途：图书馆
设计/竣工年份：2010年/2013年
建设地点：浙江省宁波市北仑区
总建筑面积：37 375 m²
总建筑高度/层数：23.85 m /5 F
结构形式：框架结构

Owner: Bureau of Construction Works, Beilun District, Ningbo City
Function: Library
Design/Completion Year: 2010/2013
Construction Site: Beilun District, Ningbo City, Zhejiang
Total Floor Area: 37,375 m²
Total Height/Floor: 23.85 m /5 F
Structure: Frame Structure

浙江档案馆新馆
Zhejiang Provincial Archives New Building

建设单位：浙江省档案馆
功能用途：档案馆
设计/竣工年份：2011年/～
建设地点：浙江省杭州市丰潭路
总建筑面积：52 800 m²
总建筑高度/层数：67.2 m/15 F
结构形式：框剪结构

Owner: Zhejiang Provincial Archives
Function: Archives
Design/Completion Year: 2011/~
Construction Site: Fengtan Road, Hangzhou City, Zhejiang
Total Floor Area: 52,800 m²
Total Height/Floor: 67.2 m/15 F
Structure: Frame-Shear Wall Structure

　　本项目的建筑设计，吸取了档案馆成立以来频繁移动馆址的教训，完善了档案馆建筑的功能需求，并留有充足的发展余地。新馆设计以"公共服务、安全保管、文化传承、以人为本"为主线，在总体布局上，做到内外有序，合理控制公共开放空间的尺度和环境景观设置，形成"既开又合、内外兼顾、适度开放"的建筑空间，并为今后的发展留有一定的余地；在建筑造型设计上，从功能出发，以神似竹简的竖向凹缝来处理，造型简洁庄重，遵循适用、经济、美观、大方的原则，体现节能环保的设计理念，与周边环境相协调。

Architectural design of this project learns the lessons in frequent relocation of the archives after foundation, improves the functional requirements of archives building, and reserves enough space for future development. The design of the new building follows a main clue of "public service, safe maintenance, cultural inheritance and human-oriented". The aim of overall layout is to realize regular internal and external spaces, appropriate control of scale of public open space and arrangement of environmental landscape, diversified open and controlled spaces, and to reserve certain space for future development; the design of building shape focuses on functional layout and boasts of bamboo-like longitudinal concave joint to create simple, solemn, practical, economical, esthetic and elegant image, to reflect the energy-saving and environment-friendly concept and to realize harmonious relation with contextual environment.

新疆和田影剧院
Xinjiang Hetian Theater

建设单位：浙江省援疆办
功能用途：剧院
设计/竣工年份：2009年/～
建设地点：新疆和田地区
总建筑面积：10 690 m²
总建筑高度/层数：24 m/5 F
结构形式：框架结构

Owner: Zhejiang Xinjiang Assistance Office
Function: Theater
Design/Completion Year: 2009/~
Construction Site: Hetian Prefecture, Xinjiang
Total Floor Area: 10,690 m²
Total Height/Floor: 24 m/5 F
Structure: Frame Structure

项目位于新疆维吾尔自治区和田市中心地段，位于迎宾路东侧，与和田市地委大楼隔路相望。影剧院设计规模为1 000座。考虑到大量的观众、车辆和演职人员的进出，剧院主要设有四个出入口，分别为观众入口、贵宾入口、演职人员入口和舞台器械货运入口，以满足人们的各种出入需求。影剧院在造型上犹如一朵几欲绽放的沙漠玫瑰，既不羞涩亦不张扬，恰到好处地诠释着和谐之美。在立面纹理上以展示和田的历史文化为基调，整体立面犹如无数条丝带迎风飞舞，舞动出一个个动人的乐章，象征着和田人民热情奔放的性格。

The project is located in the central district of Hetian Prefecture, Xinjiang, and at east of Yingbin Road, facing to the Hetian Committee Building across the road. The theater will provide 1,000 seats. Considering the large flow of audiences, vehicles, actors and employees, four main entrances are designed for audiences, VIP guests, actors and stage instruments respectively, so as to satisfy different access requirements. The theater building is like a desert rose bud, appropriately embodying the beauty of harmony. The facade is designed to exhibit the historical culture of Hetian and the whole facade is just like numerous flying ribbons, playing euphonic music and embodying the passion of Hetian people.

医院建筑
Medical Care Building

198 温州医学院附属第二医院瑶溪分院
200 天台县人民医院迁建工程
201 浙江省中医院国家中医临床研究基地科研综合楼
202 平阳县人民医院异地扩建工程
204 台州市立医院新院区
205 苍南县人民医院迁建工程
206 杭州市儿童医院医疗综合楼
208 开化县人民医院门急诊综合楼
209 浙江省新华医院门诊及急诊住院楼改造工程
210 浙江大学医学院附属义乌医院
212 东阳市人民医院医疗综合大楼
214 北仑人民医院

198 Yaoxi Branch of the Second Affiliated Hospital of Wenzhou Medical College
200 Relocation Project of Tiantai County People's Hospital
201 National Traditional Chinese Medicine Clinical Research Base Building in Zhejiang Provincial Hospital of Traditional Chinese Medicine
202 Expansion Project of Pingyang County People's Hospital
204 New Zone of Taizhou Municipal Hospital
205 Relocation Project of Cangnan County People's Hospital
206 Medical Building of Hangzhou Children's Hospital
208 Outpatient Building of Kaihua County People's Hospital
209 Outpatient Building Renovation Project in Zhejiang Xinhua Hospital
210 Affiliated Yiwu Hospital of Medicine School, Zhejiang University
212 General Medical Building of Dongyang People's Hospital
214 Beilun People's Hospital

温州医学院附属第二医院瑶溪分院
Yaoxi Branch of the Second Affiliated Hospital of Wenzhou Medical College

项目包括普通医院（500床）和儿童医院（500床）。整体设计上，我们着力体现"便捷、高效"的原则，充分按照普通成人医院和儿童医院相结合的特征，将儿童医院的设计概念贯穿设计过程，做到对儿童病人和成人病人心理特点的双重关怀。平面设计强调医患分流和洁污分流。门诊、医技、病房均设置有员工专用通道，与患者通道互不干扰，利于效率提高与资源共享。主立面设计大胆采用了彩色玻璃幕墙，不规则的玻璃色块，既清爽活泼，符合儿童心理特点，又能使立面脱颖而出，突出建筑形象。

The project is composed of a general hospital (500 beds) and a children's hospital (500 beds). In the overall design, we focus on expressing the principle of "convenience and efficiency". Fully complying with feature combined the general adult hospital and children's hospital, the design process penetrates the design concept of the children's hospital to achieve the dual care of the children patient and adult patient's psychology feature. The graphic design stresses the doctor-patient triage, and clean-sewage diversion. The clinic, medical technician and ward set the staff dedicated channel not interfering with the patient channel, thus helping improving the efficiency and sharing the resource. The main facade design boldly uses the colored glass curtain wall. The irregular glass patch not only is fresh and lively, complying with the children's psychological feature, but also highlights the facade and building image.

建 设 单 位：温州医学院附属第二医院
功 能 用 途：综合医院
设计/竣工年份：2012年/~
建 设 地 点：浙江省温州市永强片区
总建筑面积：120 550 m²
总建筑高度/层数：72 m/15 F
结 构 形 式：框剪结构

Owner: The Second Affiliated Hospital of Wenzhou Medical College
Function: General Hospital
Design/Completion Year: 2012/~
Construction Site: Yongqiang Plot, Wenzhou City, Zhejiang
Total Floor Area: 120,550 m²
Total Height/Floor: 72 m/ 15 F
Structure: Frame-Shear Wall Structure

总平面图

天台县人民医院迁建工程
Relocation Project of Tiantai County People's Hospital

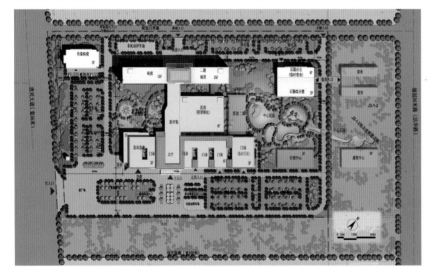

建 设 单 位：天台县人民医院
功 能 用 途：医院
设计/竣工年份：2012年/2015年
建 设 地 点：浙江省台州市天台县
总 建 筑 面 积：80 021 m²
总建筑高度/层数：61.5 m/15 F
结 构 形 式：框架结构

Owner: Tiantai People's Hospital
Function: Hospital
Design/Completion Year: 2012/2015
Construction Site: Tiantai County, Taizhou City, Zhejiang
Total Floor Area: 80,021 m²
Total Height/Floor: 61.5 m/ 15 F
Structure: Frame Structure

项目将门诊、医技、病房三大功能区设计成"王"字形，流线最短，联系便捷。急诊区位于建筑西南侧，与门诊、医技区联系紧密，并靠近主入口，便于患者到达，夜间可独立对外运营。门诊、医技科室围绕着主次医疗街排布，通过设置若干景观院落，易于患者辨识而不会迷失，利于丰富患者的行为感知。在立面造型上，建筑体量适宜，平面规整，均有利于充分利用各个空间场所；裙房和病房大楼均采用连续的直角折线所构成的整体条形窗，不仅具有刚性的岩石肌理，还有效地增加了采光面积。沿水面的建筑强调水平线条的横向延伸感。

The project displays 王-shaped relationship among the clinic, medical technician and ward, the flow line is the shortest and it is convenient for the connection. The emergency area is located at the southwest side of the building, closely related with the clinic and medical technician. It is close to the main entrance, thus facilitating the arrival of the patient. During the night, it can be operated to the outside separately. The clinic and medical technician rooms are set surrounding the primary and secondary medical street. Through setting several landscape courtyard, it is easy for the patient to recognize and not to get lost, and help enrich the patient's behavior perception. In the facade modeling, the suitable building volume and the regular plane can help fully use each spatial place. The annex and ward building adopt the continuous rectangular polyline to constitute the overall bar window, which can not only have the rigid rock texture, but also effectively add the lighting area. The building along with the water stresses the horizontal extending feeling of the horizontal line.

浙江省中医院国家中医临床研究基地科研综合楼
National Traditional Chinese Medicine Clinical Research Base Building in Zhejiang Provincial Hospital of Traditional Chinese Medicine

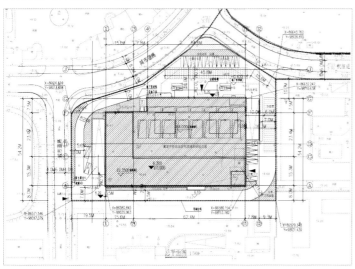

建 设 单 位：浙江省中医院
功 能 用 途：病房、科研楼
设计/竣工年份：2012年/~
建 设 地 点：浙江省杭州市下沙九号大街
总 建 筑 面 积：55 300 m²
总建筑高度/层数：99.6 m/22 F
结 构 形 式：框剪结构

Owner: Zhejiang Provincial Hospital of Traditional Chinese Medicine
Function: Ward, Scientific Research
Design/Completion Year: 2012/~
Construction Site: No.9 Avenue, Xiasha District, Hangzhou City, Zhejiang
Total Floor Area: 55,300 m²
Total Height/Floor: 99.6 m/22 F
Structure: Frame-Shear Wall Structure

本工程为国家发改委和国家中药管理局共同确定的全国16家国家中医临床研究基地建设项目之一，也是全国唯一一家重点研究血液病的研究基地。总平面布局充分考虑了新建建筑与原有医院院区之间以及与未来扩建用地之间的关系，使得三者有效结合。建筑平面布局方正，充分考虑了科研基地和扩建病房楼的合理分区。病房平面采用复廊式布局，力求做到对病人和医务人员的双重关怀。造型设计采用简约的设计手法，立面强调竖线条的明快，展现富有韵律的建筑个性，创造出一个具有时代气息、开放、大气的建筑形象。

The project is one of national 16 Chinese medicine clinical research base construction projects jointly determined by National Development and Reform Commission (NDRC) and State Administration of Traditional Chinese Medicine of the People's Republic of China (SATCM), and it is the only one key research base for the hematology. The overall layout fully considers the relationship among the new building and original hospital and the future expansion land, thus effectively combining three factors. The building layout is square, fully considering the reasonable partition of the scientific base function and expanding ward building. The ward plane adopts the double corridor layout to achieve the dual care of the patient and the medical staff. The style design adopts the simple design technique. The facade stresses the brightness of the vertical line, displays the rhythmic building feature, and creates a modern, open and ambient building image.

平阳县人民医院异地扩建工程
Expansion Project of Pingyang County People's Hospital

项目总用地面积 98 524 m², 床位1 077个, 日门诊量3 000人次。本工程包括七个子项, 分别为病房综合楼、门诊医技楼、行政综合楼、公共卫生楼、后勤保障楼、值班宿舍楼和高压氧舱。在主体医疗区布局上, 采用现代化的医疗街模式, 结合中庭、绿化庭院, 避免大型医院的迷宫式布局。围绕着开敞明亮的景观中庭, 具有生态景观的医疗街巧妙地将医疗区划分为门诊区、医技区、病房楼、行政办公区。立面以虚实对比的手法创造出大气而统一的形象, 力求打破传统医院冰冷严肃的建筑性格, 营造高情感、人性化的建筑体验。

The project covers a site area of 98,524 m² with 1,077 beds and 3,000 daily outpatients. The project has seven subitems: ward building, outpatient medical technology building, administrative building, public health building, logistics building, duty dormitory and high-pressure oxygen chamber. In the main medical area layout, it adopts the modern medical street pattern, and combines with the atrium and green courtyard to avoid the labyrinth layout in the large hospital. Surrounding the open and bright landscape atrium, the medical street with ecological landscape skillfully divides the medical area into the outpatient district, medical technology district, ward building and administrative office. The facade creates an ambient and unified image with the virtual-real comparison, it seeks to break the cold and serious building feature in the traditional hospital, and create the high-emotion and humane building experience.

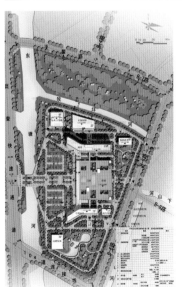

建设单位：平阳县人民医院
功能用途：综合医院
设计/竣工年份：2011年/~
建设地点：浙江省温州市平阳县
总建筑面积：121 830 m²
总建筑高度/层数：59.8 m/15 F
结构形式：框剪结构

Owner: Pingyang People's Hospital
Function: General Hospital
Design/Completion Year: 2011/~
Construction Site: Pingyang County, Whenzhou City, Zhejiang
Total Floor Area: 121,830 m²
Total Height/Floor: 59.8 m/ 15 F
Structure: Frame-Shear Wall Structure

台州市立医院新院区
New Zone of Taizhou Municipal Hospital

建设单位：台州市立医院
功能用途：综合医院
设计/竣工年份：2011年/～
建设地点：浙江省台州市椒江区
总建筑面积：197 616 m²
总建筑高度/层数：78 m/19 F
结构形式：框剪结构

Owner: Taizhou Municipal Hospital
Function: General Hospital
Design/Completion Year: 2011/～
Construction Site: Jiaojiang District, Taizhou City, Zhejiang
Total Floor Area: 197,616 m²
Total Height/Floor: 78m/19 F
Structure: Frame-Shear Wall Structure

该项目是一所集医疗、科研、教学及预防、保健于一体的综合性医院。设计基于周边环境，结合可持续发展的整体设计策略，合理布局各单体建筑。在单体建筑方面，设计通过对门急诊、医技与发展用房等不同体量进行组合，以前低后高的空间关系将建筑整合为一组大景深的图景。立面采用活泼的错格肌理设计，辅以明快的颜色点缀，弱化医疗建筑冰冷严肃的形象，为城市增添亮丽的风景。

The project is a comprehensive hospital integrating the medical treatment, research, teaching, prevention and health care. Based on the surrounding environment and combined with the overall design strategy of sustainable development, the design rationally arranges the building. In the individual building, the design combines the outpatient, medical technology and development room, integrates the building into a group of extended depth of field picture with the spatial relationship of front-lower and back-higher. The facade adopts the lively stagger grid texture added with the bright color to weaken the cold and serious image of the medical building and add a beautiful landscape for the city.

苍南县人民医院迁建工程
Relocation Project of Cangnan County People's Hospital

建设单位：苍南县人民医院
功能用途：综合医院
设计/竣工年份：2012年/~
建设地点：浙江省温州市苍南县
总建筑面积：122 657 m²
总建筑高度/层数：61.5 m/15 F
结构形式：框剪结构

Owner: Cangnan People's Hospital
Function: General Hospital
Design/Completion Year: 2012/~
Construction Site: Cangnan County, Whenzhou City, Zhejiang
Total Floor Area: 122,657 m²
Total Height/Floor: 61.5 m/15 F
Structure: Frame-Shear Wall Structure

本工程为苍南县人民医院迁建工程，是苍南县多年以来最大的一项民生工程。总体设计坚持贯彻"高起点、高标准、高水平"的原则，既有超前意识，又充分考虑到现实可能，厉行节约、节能、生态，运用有限的资金塑造现代化医疗中心。同时重视环境，结合中庭、绿化庭院，形成拥有绿色生态景观的大型综合医院。

This is the relocation project of Cangnan County People's Hospital, and it is the largest livelihood project in Cangnan County in these years. The overall design persists in implementing the principle of "high starting-point, high standard and high level", not only having the forward-thinking, but also fully considering the realistic possibility. The project is energy-saving and ecological, uses the limited fund to shape the modern medical center. Meanwhile, we shall focus on the environment, combine with the atrium and green garden to form the large comprehensive hospital with green landscape.

杭州市儿童医院医疗综合楼
Medical Building of Hangzhou Children's Hospital

项目在有限的用地范围内采用集约化、高层化模式，将主楼布置在地块西北角，采用短板与弧线结合的手法，与原大楼拉开间距并遥相呼应。门诊医技部分主要集中在病房楼1~6层，两者"U"形的布局围合成一个通高的门诊大厅，与原大楼共同围合出一个面向主入口广场的建筑形态。平面设计强调医患分流、洁污分流，以求效率提高与资源共享。通过候诊空间和医务服务空间，各种流线彼此连接，相辅相成，从而形成医疗区的主要流线骨架。

The project adopts the intensive and high-level pattern within the limited land, and arranges the main building at the northwest corner. It adopts the method combining the short slab and arc to open the space and echo with the original building. The outpatient medical technology part mainly centres on 1-6 layers of the ward building. The U-shaped layout constitutes a floor-to-ceiling outpatient hall, and surrounds a building form facing the main entrance square with the original building. The graphic design stresses the doctor-patient triage, clean-sewage diversion to seek the improvement of the efficiency and the sharing of resource. Through the waiting space and medical service space, all flow lines are connected, thus forming main skeleton in the medical area.

建设单位：杭州市儿童医院
功能用途：医疗
设计/竣工年份：2011年/~
建设地点：浙江省杭州市上塘路
总建筑面积：60 078 m²
总建筑高度/层数：69 m/17 F
结构形式：框架结构

Owner: Hangzhou Children's Hospital
Function: Medical Treatment
Design/Completion Year: 2011/~
Construction Site: Shangtang Road, Hangzhou City, Zhejiang
Total Floor Area: 60,078 m²
Total Height/Floor: 69 m/ 17 F
Structure: Frame Structure

开化县人民医院门急诊综合楼
Outpatient Building of Kaihua County People's Hospital

建 设 单 位：开化县人民医院
功 能 用 途：医院
设计/竣工年份：2010年/2013年
建 设 地 点：浙江省衢州市开化县
总 建 筑 面 积：15 602 m²
总建筑高度/层数：23.85 m/6 F
结 构 形 式：框架结构

Owner: Kaihua People's Hospital
Function: Hospital
Design/Completion Year: 2010/2013
Construction Site: Kaihua County, Quzhou City, Zhejiang
Total Floor Area: 15,602 m²
Total Height/Floor: 23.85 m/ 6 F
Structure: Frame Structure

项目位于开化县现人民医院用地内，为原有门急诊楼的拆除重建项目。建筑西南两面临城市道路，北侧为医院的住院楼，东侧为医院职工的生活区。新建门急诊综合楼为集中式设置，通过建筑中部一条东西向的公共医疗通廊，对流线进行合理组织。建筑外形坚持"形式追随功能"，在布局良好的基础上，强调建筑的时代感和稳重感，建筑立面肌理强调秩序性的变化，使建筑在简洁的线条下蕴含丰富的内层肌理。该项目为一座功能和形式相互呼应，造型和细部相互衬托的现代化医疗建筑。

The project is located at the existing Kaihua County People's Hospital, and it is the demolition and reconstruction project of the original outpatient building. The western and southern part of the building face the urban road, the northern side is the inpatient building, and the western side is the living area for the hospital staff. The newly-built outpatient building is the centralized settlement. Through the east-west public health corridor at the middle of the building, the flow line is rationally organized. The building's appearance adheres to the "form following function". Based on the good layout, it stresses the building's modernity and sedateness. The building facade texture stresses the change of the order, thus making the building contain a wealth of inner texture under the simple lines, echoing the function and form, and it is a modern medical building mutually foiling the shape and detail.

浙江省新华医院门诊及急诊住院楼改造工程
Outpatient Building Renovation Project in Zhejiang Xinhua Hospital

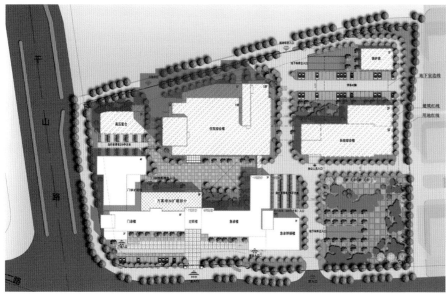

建 设 单 位：浙江省新华医院
功 能 用 途：门诊、急诊
设计/竣工年份：2012年/~
建 设 地 点：浙江省新华医院内
总建筑面积：17 030 m²
总建筑高度/层数：24 m/7 F
结 构 形 式：框架结构

Owner: Zhejiang Xinhua Hospital
Function: Clinic, Emergency
Design/Completion Year: 2012/~
Construction Site: Zhejiang Xinhua Hospital
Total Floor Area: 17,030 m²
Total Height/Floor: 24 m/ 7 F
Structure: Frame Structure

现有门诊楼建造于1986年，6层，是较老的建筑。现有的急诊住院楼建造于1995年，为7层的多层建筑，门诊楼及急诊住院楼设计使用年限为50年。我们在尊重原有设计和结构安全性的基础上，在满足外墙保温隔热要求的前提下，不做大的改动，并且力求和东侧的急诊住院楼已完成干挂的立面相协调。外墙保留原有设计中的小雨棚和遮阳板，色调尽可能与急诊楼外墙色彩统一。在室内主要的公共空间延续原有的建筑风格，塑造内外交融、通透坚固的空间特质。

The existing outpatient building was built in 1986, and it is a six-storey old building. The existing emergency building was built in 1995, and it is a seven-storey building. The design working life of the building is 50 years. Based on the respect of the original design and structural safety and under the premise of meeting the requirement of external wall insulation, no more large changes are conducted. We seek to make the building coordinate with the dry-hanging facade of the eastern emergency building. The outer wall retains the small canopy and sun louver in the original design, and the hue shall be unified with the exterior wall of the emergency building as much as possible. Main indoor public space continues the original building style, thus shaping the spatial quality of internal and external blending, transparent and solid.

浙江大学医学院附属义乌医院
Affiliated Yiwu Hospital of Medicine School, Zhejiang University

项目设计体现"以病人为中心"的设计理念。利用基地与南侧商城大道的高差，医院的主入口处设置架空的医院前广场，通过前广场的立体交通实行严格的人车分流，保证行人安全及车流快速疏导。门诊、医技、病房三个区呈等边三角形关系，流线最短，联系便捷。急诊区位于病房下面，与门诊区、医技区、病房区关系紧密。医院的医疗主通道围绕景观内院布置，易于患者亲近自然而又不会迷失方向；医疗大楼设计为患者走中间阳光医疗街，医生走两侧专用通道。医院内部的垂直和水平交通均能保证医患流线完全分开。

The project reflects the design concept of "patient-centered". With the altitude difference between the base and the Shangcheng Avenve in the south, the main entrance of the hospital sets the overhead hospital square. Through the three-dimensional transportation in the square, it implements the strict pedestrian and vehicle separation to ensure the rapid persuasion. The clinic, medical technology building and ward display the equilateral triangle relationship, the streamline is the shortest and it is convenient to connect. The emergency area is set below the ward, closely related with the clinic, medical technology building and ward. The medical main channel of the hospital is set around the landscape courtyard, it is convenient for patients to go into the natural environment without getting lost; concerning the medical building plane, patients pursue the middle sunlight medical street, and doctors pursue the dedicated channel at two sides. The internal vertical and horizontal traffic in the hospital can ensure the total separation of the flow line of the doctor and patient.

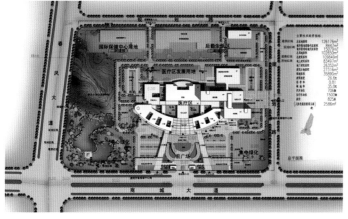

医疗大楼五~十二层平面图

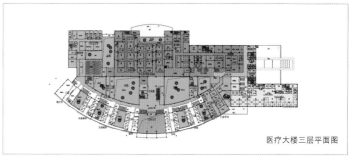

医疗大楼三层平面图

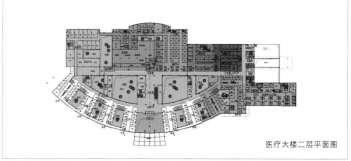

医疗大楼二层平面图

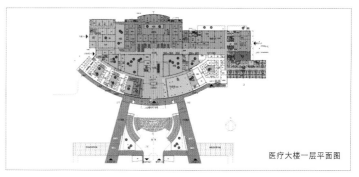

医疗大楼一层平面图

建 设 单 位：浙江大学医学院附属义乌医院
功 能 用 途：综合医院
设计/竣工年份：2009年/~
建 设 地 点：浙江省义乌市
总 建 筑 面 积：110 000 m²
总建筑高度/层数：60.75 m/15 F
结 构 形 式：框架结构

Owner: Affiliated Yiwu Hospital of Medicine School, Zhejiang University
Function: General Hospital
Design/Completion Year: 2009/~
Construction Site: Yiwu City, Zhejiang
Total Floor Area: 110,000 m²
Total Height/Floor: 60.75 m/ 15 F
Structure: Frame Structure

东阳市人民医院医疗综合大楼
General Medical Building of Dongyang People's Hospital

工程地处市中心，19层的病房楼布置在用地北侧，与现病房大楼拉开距离。急救区在西侧靠前广场布置，保证急诊流线的便捷；医技楼布置在整个医院的中心位置，能最好地被门诊、急诊、病房共享；病房楼与病房大楼、肿瘤病区形成医院完整的病房区。原急救中心底层作架空处理，使南侧广场与吴宁西路视线贯通，营造出医院开阔完整的广场空间，快速疏导门诊、急诊、病房等的人流。大楼西北侧留出15m的道路空间，形成院内快速就诊通道，便于南北双向人流、车流的疏散。大楼底层中部作架空处理，形成院内便捷的步行交通系统，将医院各医疗建筑串联成一体。

The project is located in the city center. The 19-storey ward building is arranged at the northern side, which is far away from the existing ward building. The emergency section is set in the square at the western side to ensure the convenience of emergency streamline; the medical technical department is located at the central part of the whole hospital, which can be better shared by outpatient, emergency and ward; the layout of the ward building, the existing ward building and the tumor ward can form the complete ward area in the hospital. The ground floor of the original emergency center is lifted up to connect the sight between the southern square and Wuning West Road, thus creating the open and complete square space in the hospital and rapidly diverting the flow of the outpatient, emergency and ward. 15 m road space is left at the northwest side of the building to form the quick visit channel in the hospital for the convenience of the south-north two-way traffic. The middle bottom of the building is overhead processed, thus forming the convenient walk transportation and connecting each medical building.

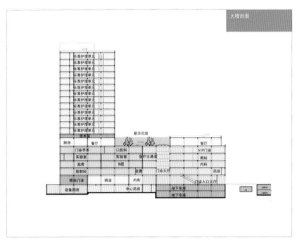

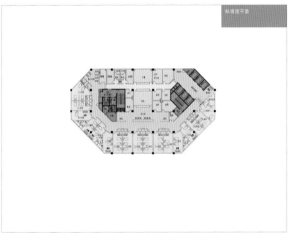
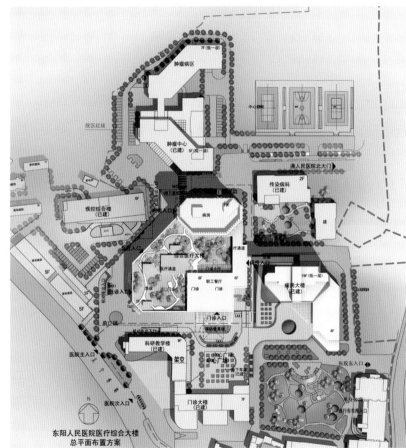
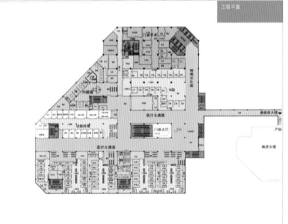
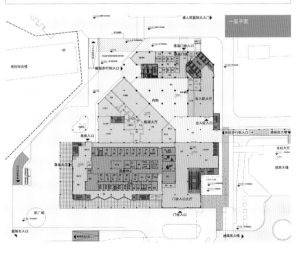

建设单位：东阳市人民医院
功能用途：门诊、急诊、医技、病房等
设计/竣工年份：2009年/2013年
建设地点：浙江省东阳市
总建筑面积：83 000 m²
总建筑高度/层数：76.4 m/19 F
结构形式：框剪结构

Owner: Dongyang People's Hosptial
Function: Clinic, Emergency, Medical Technology, Ward, etc.
Design/Completion Year: 2009/2013
Construction Site: Dongyang City, Zhejiang
Total Floor Area: 83,000 m²
Total Height/Floor: 76.4 m/19 F
Structure: Frame-Shear Wall Structure

北仑人民医院
Beilun People's Hospital

项目位于宁波市北仑区，定位为800床的三级乙等综合性医院，分为门急诊医技病房综合楼、感染门诊及病房楼、行政楼、专家公寓及宿舍楼，项目总用地面积为75 858 m²。设计采用半集中式医院规划形态和高层与多层双向发展模式，利用医疗主街连接各个医疗功能模块，布局紧凑，分区明确，流线清晰，既强调医疗效率，又注重医疗环境质量，积极利用自然通风采光，节能降耗。

The project, located in Beilun District, Ningbo City, is positioned as a Grade-III Class-B comprehensive hospital with 800 beds, which can be divided into outpatient and emergency medical technology building, infection clinic, inpatient building, administration building, expert apartment and dormitory. It covers a site area of 75,858 m². The design adopts the semi-centralized hospital planning form, high-rise and multi-layer bidirectional development mode, and uses the medical main street to connect each medical functional module. The layout is compact, the partition is clear, and the flow line is clear. It not only stresses the medical efficiency, but also focuses on the medical environment quality. It actively uses the natural ventilation and lighting, which can save the energy and reduce the consumption.

建 设 单 位：北仑区建筑工务局
功 能 用 途：医疗
设计/竣工年份：2008年/2011年
建 设 地 点：浙江省宁波市北仑区
总建筑面积：102 000 m²
总建筑高度/层数：49.35 m/12 F
结 构 形 式：框剪结构
合 作 单 位：上海卫生建筑设计研究院有限公司

Owner: Bureau of Construction Works Beilun District
Function: Medical Treatment
Design/Completion Year: 2008/2011
Construction Site: Beilun District, Ningbo City, Zhejiang
Total Floor Area: 102,000m²
Total Height/Floor: 49.35m/12F
Structure: Frame-Shear Wall Structure
Cooperation Unit: Shanghai Health Construction Design Institute Co., Ltd.

住宅建筑
Residential Building

218	中铁·青秀城
220	湖北荆门双仙社区
222	临沂"联合·世纪新筑"B地块
223	中大·杭州西郊半岛妙得阁
224	黄岩东路安置房
226	玉环县峦岩山小区
227	山东临沂"旭徽·凤凰水城"
228	杭州萧山潮闻天下
230	库尔勒棉纺厂地块
232	衢州保障性住房——锦西家园
233	温岭利兹·水印华庭
234	江干区笕桥黎明社区拆迁安置房项目
236	荆门市凯凌·香格里拉小区三期
238	贵阳市花溪区十和田项目
240	黄岩羽村安置房工程

218	Zhongtie • Qingxiu City
220	Shuangxian Community in Jingmen, Hubei
222	Linyi "Lianhe • Excellent Life" Plot B
223	Zhongda • Hangzhou West Suburb Peninsula Miaode Pavilion
224	Huangyan Donglu Resettlement Housing Project
226	Luanyan Mountain Residential District, Yuhuan County
227	"Xuhui • Lake of Phoenix" in Linyi, Shandong
228	Chaowen Tianxia in Xiaoshan, Hangzhou
230	Korla Cotton Mill Lot
232	Quzhou Indemnificatory Housing—Jinxi Garden
233	Wenling Lizi • Shuiyin Garden
234	Resettlement Houses of Liming Community in Jianqiao, Jiangkan District
236	Jingmen Kailing • Shangri-La Phase III Project
238	Guiyang Huaxi Shihetian Project
240	Huangyan Yucun Village Resettlement Housing Project

242	江苏汇丰城市综合体方案
243	山东临沂"爱伦坡"
244	杭州寰宇天下
246	杭长铁路等重点工程农居拆迁安置房
248	三门银轮玫瑰湾小区
249	芜湖县南湖品园小区
250	杭州阳光郡
252	萧山浪琴湾
254	山东淄博"正承·PARK"
256	恒隆国际花园及恒隆国际酒店
257	萧山云涛名苑
258	山东郯城中央华庭
260	当涂豪邦君悦华庭住宅小区
261	浙江天台红石梁广场

242	Jiangsu HSBC Urban Complex
243	"Ailunpo" in Linyi, Shandong
244	Hangzhou U World
246	Resettlement Housing for Farmers for Major Projects like Hangzhou-Changsha Passenger-dedicated Line
248	Sanmen Yinlun Rose Bay Residential District
249	Nanhu Pinyuan Residential District in Wuhu County
250	Hangzhou Sunny Estate
252	Xiaoshan Langqin Bay
254	"Zhengcheng • PARK" in Zibo, Shandong
256	Henglong International Garden and Henglong International Hotel
257	Xiaoshan Yuntao Garden
258	Central Garden in Tancheng, Shandong
260	Dangtu Haobang Junyuehuating Residential District
261	Zhejiang Tiantai Hongshiliang Square

中铁·青秀城
Zhongtie·Qingxiu City

项目位于萧山区建设一路与明星路交会处，由12幢18层的统一朝向的高层住宅及沿街住宅服务用房组成。利用地块方形的特性，总体采用轴对称布局。东西两侧布置点式住宅，南侧和北侧布置板式住宅。5#、6#楼底层架空，前后中心花园相互衔接，打造出一个20 000 m²的中央花园。景观设计为新古典欧式轴线对称与曲线自然混搭风格。对中央庭院进行合理布局，形成以"水"为主题的中央水景区、以"舞"为主题的运动健身区以及规则的大草坪区。外立面设计采用新古典风格，给人一种高贵雅致、舒适高档的居住感受。

The project, located at the junction of 1st Jianshe Road and Mingxing Road in Xiaoshan District, is composed of twelve 18-storey high-rise buildings with the same orientation and residential service houses along the street. Based on the square shape of the lot, axial symmetry is adopted for the overall arrangement. Point-type residence is adopted on the east and west sides, while slab-type residence is adopted on the north and south sides. The ground floors of Building 5# and 6# are built on stilts, and the front and back central gardens are linked up, to create a 20,000 m² central garden. The landscape design is natural hybrid style of neoclassical European axial symmetry and curves. The rational layout of the central courtyard forms a central waterscape themed by "water", a fitness zone themed by "dance" and large formal lawns. The facade is of neoclassical style, which looks elegant, comfortable and upscale.

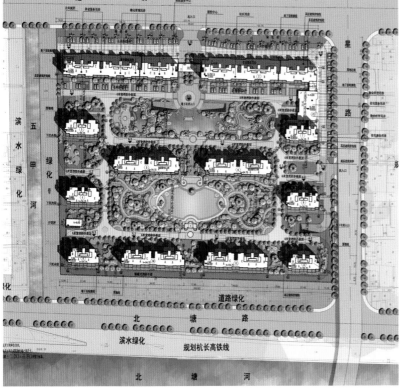

建设单位：中铁房地产集团杭州京顺置业有限公司
功能用途：住宅、商业
设计/竣工年份：2013年/2015年
建设地点：浙江省杭州市萧山区
总建筑面积：165 000 m²
总建筑高度/层数：60 m/18 F
结构形式：框剪结构

Owner: Zhongtie Real Estate Group Hangzhou Jingshun Real Estate Co., Ltd.
Function: Residence, Commerce
Design/Completion Year: 2013/2015
Construction Site: Xiaoshan District, Hangzhou City, Zhejiang
Total Floor Area: 165,000 m²
Total Height/Floor: 60 m/18 F
Structure: Frame-Shear Wall Structure

湖北荆门双仙社区
Shuangxian Community in Jingmen, Hubei

项目用地位于荆门市漳河新区双溪街道,根据用地性质划分为居住用地和商业办公用地两个地块,分别位于用地的东侧和西北侧。住宅地块划分为A、B、C共三个组团,高层住宅共21栋;高层商业综合体位于住宅地块西北角;物业管理用房分别位于1#、15#楼底层;社区活动中心位于1#楼底层;幼儿园位于住宅地块东北角。

各住宅组团的绿化围合成一个带形的中心景观,与小区沿城市道路的绿化带相互渗透。单体建筑以深色涂料、玻璃作为主要外墙材料,色彩上以深咖啡色、偏暖色调为主,塑造安宁温馨的高档住宅社区。

The project is located in Shuangxi Subdistrict, Zhanghe New District, Jingmen City, Hubei Province. According to the land usage, the project land is divided into two lots—commercial lot and residential lot, which are located at the east and northwest sides of the project land respectively. The residential lot is divided into three building clusters—Cluster A, B and C. There are a total of 21 residential high-rise buildings. The commercial high-rise complex is located in the northwest corner of the residential lot; the property management houses are located in the ground floors of Building 1# and 15#; the community activity center is located in the ground floor of Building 1#; and the kindergarten is located in the northeast corner of the residential lot.

The green areas of the building clusters form a belt central landscape, which is connected to the green belt of the residential district along the urban road. The outer walls of the individual buildings are mainly featured by glasses and dark paint, mostly dark coffee and warm colors, to create a peaceful and warm living environment as high-end residential buildings.

建设单位：湖北荆门城建集团房地产开发有限公司
功能用途：住宅
设计/竣工年份：2013年/～
建设地点：湖北省荆门市
总建筑面积：143 823 m²
总建筑高度/层数：53.65 m /18 F
结构形式：框剪结构

Owner: Hubei Jingmen Urban Construction Group Real Estate Development Co., Ltd.
Function: Residence
Design/Completion Year: 2013/～
Construction Site: Jingmen City, Hubei
Total Floor Area: 143,823 m²
Total Height/Floor: 53.65 m /18 F
Structure: Frame-Shear Wall Structure

临沂"联合·世纪新筑"B地块
Linyi "Lianhe · Excellent Life" Plot B

建设单位：临沂新筑置业有限公司
功能用途：住宅、商业
设计/竣工年份：2013年/～
建设地点：山东省临沂市北城新区
总建筑面积：77 000 m²
总建筑高度/层数：24 m/7 F
结构形式：框剪结构

Owner: Linyi Xinzhu Real Estate Co., Ltd.
Function: Residence, Commerce
Design/Completion Year: 2013/~
Construction Site: Beicheng New District, Linyi City, Shandong
Total Floor Area: 77,000 m²
Total Height/Floor: 24 m/7 F
Structure: Frame-Shear Wall Structure

项目位于临沂市北城新区三河口核心地带，为"联合·世纪新筑"项目的三期工程，用地面积42 461 m²，由11幢7层的花园洋房、1幢6层的综合楼及1个幼儿园组成，总建筑面积约78 000 m²。

项目从整体规划到单体细部处理都力求体现一个"精"字，精益求精、精雕细琢，耐人寻味。

建筑风格采用现代简欧主义，营造典雅、自然、高贵的气质和浪漫的情调是本案的主题。

The project, located in the central area of Sanhekou in Beicheng New District, Linyi City, is Phase III of "Lianhe · Excellent Life", covering a total site area of 42,461 m². It consists of eleven 7-storey garden houses, one 6-storey general building and one kindergarten.
Efforts are made to realize "elegance" in both overall layout and detail design.
Based on modern European-style minimalism, this project focuses on the theme of elegant, natural and noble appearance and romantic atmosphere.

中大·杭州西郊半岛妙得阁
Zhongda·Hangzhou West Suburb Peninsula Miaode Pavilion

建 设 单 位：富阳中大房地产有限公司
功 能 用 途：住宅
设计/竣工年份：2013年/～
建 设 地 点：浙江省杭州市富阳
总 建 筑 面 积：105 439 m²
总建筑高度/层数：98.6 m/33 F
结 构 形 式：框架结构
合 作 单 位：美国JWDA建筑设计事务所
　　　　　　　上海骏地建筑设计咨询有限公司

Owner: Fuyang Zhongda Real Estate Co., Ltd.
Function: Residence
Design/Completion Year: 2013/~
Construction Site: Fuyang, Hangzhou City, Zhejiang
Total Floor Area: 105,439 m²
Total Height/Floor: 98.6 m/33 F
Structure: Frame Structure
Cooperation Unit: JWDA Architectural Design Firm/
Shanghai Jundi Architectural Design and Consultation
Co,. Ltd.

项目位于杭州富阳富春江畔东大道北侧，南侧紧临东大道，并与富春江和生态岛相望。地块内建筑沿南向根据地形及道路特点，点板结合，形成活跃的空间形态组合，并与周边原有建筑风格相协调，形成连续的建筑景观界面。立面设计采用现代的建筑语言，强调内外空间的渗透及层次变化，突出建筑的大气及整体性。主体建筑采用深、浅两种颜色的仿石涂料搭配，突出建筑体量的变化及形体的穿插。建筑底部使用了部分深色石材，增加体量的稳重及大气。

The project, located at the north of East Boulevard by Fuchun River in Fuyang County, Hangzhou City, is adjacent to the East Boulevard at the south and faces to the Fuchun River and Ecological Island. The buildings are planned to face the south according to the terrain and road conditions. Point-type and slab-type buildings are planned with certain angle from each other, forming vigorous spatial layout, responding to contextual existing buildings and producing a continuous building landscape. Facades are designed by using modern architectural language to highlight penetration and variation between indoor and outdoor spaces and realize generous and integral building image. The main building body decorated by stone-like coating in dark and light colors boasts of diversified and crossed volumes. The base is partially decorated with dark stone material to produce a stable and generous appearance.

黄岩东路安置房
Huangyan Donglu Resettlement Housing Project

本项目位于台州市黄岩区，总用地面积 79 652 m²，规划建设用地面积为 69 723 m²。

小区整体建筑形体以南低北高布局，内部以18层的高层住宅为主。建筑整体呈行列式布置，并且通过景观及绿化的布置将小区连接为一个有机的整体。通过中心绿地以及次级院落绿地的合理组织，考虑北侧河岸景观的共享与渗透，力求做到每一户都能享受优美的绿化景观。

The project is located in Huangyan District, Taizhou City, with a total site area of 79,652 m² and a planning construction area of 69,723 m².

The buildings in the residential district are arranged in such a manner that those in the south are lower while those in the north are higher, and most of the buildings are 18-storey high-rise buildings. The buildings are arranged in lines and are connected into an integral whole through landscape and green areas. Additionally, the design seeks to provide every house with a beautiful landscape view through the rational arrangement of the central green land and secondary green yards and the sharing and infiltration with the landscape on the river bank in the north side of the residential district.

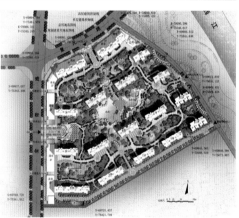

建设单位：台州市黄岩区商业街区开发建设指挥部
功能用途：住宅
设计/竣工年份：2013年/～
建设地点：浙江省台州市黄岩区
总建筑面积：224 920 m²
总建筑高度/层数：75 m / 26 F
结构形式：框剪结构

Owner: Taizhou Huangyan District Commercial Street Development and Construction Office
Function: Residence
Design/Completion Year: 2013/～
Construction Site: Huangyan District, Taizhou, Zhejiang
Total Floor Area: 224,920m²
Total Height/Floor: 75m/26F
Structure: Frame-Shear Wall Structure

玉环县峦岩山小区
Luanyan Mountain Residential District, Yuhuan County

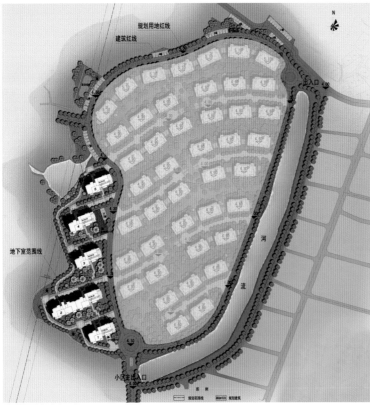

建 设 单 位：玉环地之韵房地产开发股份有限公司
功 能 用 途：住宅
设计/竣工年份：2013年/～
建 设 地 点：浙江省台州市玉环县
总 建 筑 面 积：59 271 m²
总建筑高度/层数：35 m/11 F
结 构 形 式：框剪结构

Owner: Yuhuan Dizhiyun Real Estate Development Co., Ltd.
Function: Residence
Design/Completion Year: 2013/～
Construction Site: Yuhuan County, Taizhou City, Zhejiang
Total Floor Area: 59,271 m²
Total Height/Floor: 35 m/11 F
Structure: Frame-Shear Wall Structure

本规划以尊重环境、崇尚自然、创造优美人居空间为指导思想。小区以低层住宅为主，沿西侧山体规划有少量小高层，规划人口规模为340户，1 088人。

设计着重利用基地内外景观条件进行布局，着眼于城镇大范围，整合周边优势资源，注重住宅区内外部环境的充分利用，强调住宅景观的均好性，强调空间界面的丰富性，体现以人为本的设计理念。规划布局强调视线、日照、通风，并讲究建筑的序列与肌理，在有限的用地上创造典雅的社区和建筑形象。

The planning is based on the guiding idea of respecting the environment and the nature and creating a beautiful living environment. Most of the buildings in the residential district are low-rise buildings, and there are only a few high-rise buildings along the mountain the west side. The planning population size is 340 households and 1,088 people.
The design lays emphasis on the layout of internal and external landscape, looks at the whole town, integrates advantageous resources in the surroundings, makes full use of internal and external environments of the residential district and emphasizes the sharing of landscape as well as the richness of spatial interface. The design presents the human-oriented design concept, while the planning layout emphasizes sight, sunshine and ventilation and attaches importance to the order and texture of the buildings, seeking to create an elegant community and building image on a limited land.

山东临沂 "旭徽·凤凰水城"
"Xuhui · Lake of Phoenix" in Linyi, Shandong

建 设 单 位：临沂旭徽置业有限公司
功 能 用 途：住宅、商业
设计/竣工年份：2012年/～
建 设 地 点：山东省临沂市北城新区
总 建 筑 面 积：673 000 m²
总建筑高度/层数：18~100 m/6~32 F
结 构 形 式：框剪结构

Owner: Linyi Xuhui Real Estate Co., Ltd.
Function: Commerce, Residence
Design/Completion Year: 2012/~
Construction Site: Beicheng New District, Linyi City, Shandong
Total Floor Area: 673,000 m²
Total Height/Floor: 18-100 m/6-32 F
Structure: Frame-Shear Wall Structure

项目地块位于山东临沂北城新区，东至沭河路，北至武汉路，南至成都路，用地面积143 804 m²，容积率3.80，绿地率35%。

整个小区是由多层住宅、小高层住宅及高层住宅组成的综合性社区，地块中心有占地面积50 000 m²的景观湖面，自然景观资源丰富。

The project, located in Beicheng New District, Linyi City, Shandong Province, adjoins Muhe Road in the east, Wuhan Road in the north and Chengdu Road in the south, with a site area of 143,804 m², a plot ratio of 3.80 and a greening rate of 35%.
The whole residential district is a comprehensive community composed of multi-storey residential buildings, middle-height residential buildings and high-rise residential buildings. There is a 50000 m² landscape lake in the center of the lot, showing abundant landscape resources.

杭州萧山潮闻天下
Chaowen Tianxia in Xiaoshan, Hangzhou

本项目主要包括规划地块R2和C3。

C3地块位于南阳镇钱塘江东侧白虎山，用地性质为居住用地，规划总用地面积为59 443 m²。

R2地块位于河庄街道三联村，用地性质为商住及商业办公用地，其中商住用地规划用地面积为131 417.3 m²，商业办公用地规划用地面积为13 333.3 m²。

建筑采用现代新亚洲风格，延续地域独有的历史文脉。采用深灰色的缓坡顶、深远的大挑檐、厚重的石材、高档的真石漆、深灰色金属压顶收边、局部深红色的木格栅，以现代元素重构中式建筑的神韵。

The project mainly includes planning Lot R2 and Lot C3.

Lot C3, located in Baihu Mountain on the east side of the Qiantang River in Nanyang Town, is used to build residential houses, and its planning land area is 59,443 m².

Lot R2, located in Sanlian Village, Hezhuang Subdistrict, is used to build commercial-living and office buildings, of which the planning land area for commercial residence is 131,417.3 m² and that for office is 13,333.3 m²

The modern New Asian style is adopted, to extend the unique regional historical content. Dark grey low-pitched roofs, far-reaching large overhanging eaves, heavy stones, high-end paint, dark grey metal sides and local crimson wooden grate are adopted. Modern elements are used to reinterpret the charm of Chinese buildings.

建 设 单 位：杭州百合房地产开发有限公司
功 能 用 途：住宅、商业
设计/竣工年份：2012年/～
建 设 地 点：杭州市萧山区观潮城
总 建 筑 面 积：367 000 m²
总建筑高度/层数：70 m /24 F
结 构 形 式：框剪结构
合 作 单 位：上海睿风建筑设计咨询有限公司

Owner: Hangzhou Baihe Real Estate Development Co., Ltd.
Function: Residence, Commerce
Design/Completion Year: 2012/～
Construction Site: Guanchao Town, Xiaoshan District, Hangzhou City, Zhejiang
Total Floor Area: 367,000 m²
Total Height/Floor: 70 m /24 F
Structure: Frame-Shear Wall Structure
Cooperation Unit: Shanghai Ruifeng Architectural Design and Consultation Co., Ltd.

库尔勒棉纺厂地块
Korla Cotton Mill Lot

项目位于库尔勒市老城区与开发区之间，总体为"一环两心四片多组团"的规划结构，整体空间形体呈现东、西、北三侧外围高，南侧开敞，中间低的环抱之势。规划采用组团式的空间布局模式，建筑排布注重差异性与均好性的结合。

设计通过层层推进和归属感强的空间体验，打造一个典雅尊贵的高品质居所；通过人车分流的宜人环境，打造一个休闲舒适的慢生活区；通过联系紧密的邻里空间，打造一个情意浓浓的亲和家园；通过分布四周的商业和配套公建，打造一个便捷完善的社区配套网络。

The project is located between the old urban area and development area of Korla City. In an overall planning structure of "one ring, two centers, four districts and multiple clusters" and the overall spatial form, the east, west and north peripheries are high, the south side is open and the middle is low. The cluster-type spatial pattern is adopted in the planning. Emphasis is laid on the integration of difference and sharing in building layout design.

The design aims to create an elegant and high quality living dwelling district with the spatial experience emphasizing the sense of belonging, a leisure and comfortable slow living space with a pleasant environment with separated people stream and vehicle stream, a warm home with a close neighbors' space, and a convenient community complete with supporting facilities through the construction of commercial and public facilities in the surroundings.

建 设 单 位：库尔勒雅居房地产开发有限公司
功 能 用 途：住宅、商业
设计/竣工年份：2012年/～
建 设 地 点：新疆库尔勒市
总建筑面积：446 700 m²
总建筑高度/层数：100 m / 30 F
结 构 形 式：框架结构

Owner: Korla Yaju Real Estate Development Co., Ltd.
Function: Residence, Commerce
Design/Completion Year: 2012/~
Construction Site: Korla City, Xinjiang
Total Floor Area: 446,700 m²
Total Height/Floor: 100 m / 30 F
Structure: Frame Structure

衢州保障性住房——锦西家园
Quzhou Indemnificatory Housing—Jinxi Garden

建 设 单 位：衢州市政府投资项目建设中心
功 能 用 途：住宅
设计/竣工年份：2012年/～
建 设 地 点：浙江省衢州市
总 建 筑 面 积：126 183 m²
总建筑高度/层数：19.93 m/6 F
结 构 形 式：框架结构

Owner: Quzhou Governmental Investment Project Construction Center
Function: Residence
Design/Completion Year: 2012/~
Construction Site: Quzhou City, Zhejiang
Total Floor Area: 126,183 m²
Total Height/Floor: 19.93 m/6 F
Structure: Frame Structure

项目位于衢州市锦西大道以西、工程技术学校以南地块，用地面积76 786m²，拟建设多层公寓式住宅组成的住宅小区。

小区规划着力把握总体控制，确定将现代融于自然，可持续发展成为设计指导思想，精心处理人、建筑、环境三者之间的关系，以建设生态居住环境为规划目标，满足住宅的居住性、舒适性、安全性、耐久性和经济性的要求，创造一个布局合理、交通便捷、环境优美的现代化保障小区。

The project is located on the west side of Jinxi Avenue and on the south side of Quzhou Engineering Technology School in Quzhou City, with a land area of 76,786m². It is planned to build a residential district composed of multi-storey apartment residences.
In the planning of the residential district, efforts are made to take overall control, handle the relationship between human, building and environment carefully with the design idea of integrating modernism with the nature and sustainable development and the goal of building an ecological living environment, meet the requirement of building livable, comfortable, safe, durable and economical residences, and create a modern indemnificatory housing district with a rational layout, convenient transportation and a beautiful environment.

温岭利兹·水印华庭
Wenling Lizi · Shuiyin Garden

建设单位：温岭利兹房地产开发有限公司
功能用途：住宅
设计/竣工年份：2012年/~
建设地点：浙江省台州市温岭县
总建筑面积：51 439 m²
总建筑高度/层数：68.7 m /24 F
结构形式：框剪结构

Owner: Wenling Lizi Real Estate Development Co., Ltd.
Function: Residence
Design/Completion Year: 2012/~
Construction Site: Wenling County, Taizhou City, Zhejiang
Total Floor Area: 51,439 m²
Total Height/Floor: 68.7 m /24 F
Structure: Frame-Shear Wall Structure

　　本项目用地东至TP030315地块，南至东湖路，西至9 m宽的规划道路，北至北山支河。
　　小区绿化利用四个建筑单体围合成一个L形的中心景观，并与北侧沿河绿化带互相渗透，沿街商业单体屋顶则设计了屋顶花园，景观设计主要以多层次立体景观为主题，充分利用小区内围合集中景观的同时，结合屋顶绿化、硬地广场、休闲小品打造小区特有的精品景观，营造温馨、现代、生态的立体景观庭院。

The project lot extends to Lot TP030315 in the east, Donghu Road in the south, 9m planning road in the west and Zhihe, Beishan in the north.
The green areas between 4 individual buildings form an L-shaped central landscape, and echo with the green belt along the river in the north. A roof garden is designed on the roof of the business building along the street. In landscape design, multi-layer three-dimensional landscape is the subject. Special views combined with roof greening, hard-ground square and leisure places are created while making full use of the centralized landscape inside the residential district.

江干区笕桥黎明社区拆迁安置房
Resettlement Houses of Liming Community in Jianqiao, Jianggan District

项目位于杭州市江干区笕桥。整体规划充分利用地形，建筑布局尽量坐北朝南，争取住宅的良好朝向与景观。全区设计有多种组合方式，住宅底层架空。根据不同的建筑需求，利用道路和景观区分内外不同的层次关系，强调各居住组团内部互不干扰，保证了幽静舒适的居住氛围。

立面造型设计利用传统的建筑语言，采用现代的表现手法，通过几何雕塑感的塑造，表达传统文化内涵与现代气息和谐统一。人性化地考虑空调机摆放，使其成为整个建筑的一部分。在造型上强调多元与统一，在变化中寻求协调。将生态绿化融于其中，以增加建筑的生动性。

The project is located in Jianqiao, Jianggan District, Hangzhou City. The overall planning makes full use of the terrain. In building layout, the buildings are arranged to face south as far as possible for better orientation and views. In the design of the housing estate, there are several combinations. The ground floor of the residence is built on stilts. According to different building requirements, the design emphasizes the non-interfering space between different housing clusters by making use of roads and landscape zones, which ensures a quiet and comfortable living environment.

In facade design, traditional architectural language and modern techniques of expression are used. The geometric sculptures represent a harmonious unity of traditional culture and modern flavor. The human-oriented design for the placement of air conditioners makes them be part of the building. In modeling, diversity and unity are emphasized, which is to seek coordination in changes. Ecological afforestation is integrated in the building, which enhances the vitality of the building.

建 设 单 位：杭州市江干区农居建设管理中心
功 能 用 途：住宅
设计/竣工年份：2012年/～
建 设 地 点：浙江省杭州市江干区
总 建 筑 面 积：263 600 m²
总建筑高度/层数：50 m /15 F
结 构 形 式：框架结构

Owner: Jianggan Countryside Houses Construction and Management Center
Function: Residence
Design/Completion Year: 2012/~
Construction Site: Jianggan District, Hangzhou City, Zhejiang
Total Floor Area: 263,600 m²
Total Height/Floor: 50 m/15 F
Structure: Frame Structure

荆门市凯凌·香格里拉小区三期
Jingmen Kailing・Shangri-La Phase III Project

本项目位于荆门市天鹅广场西侧的西宝山地块。基地以丘陵坡地为主,整个地形向东逐步平缓并在东面四干渠水系处终结。

设计力图使建筑成为一种融合了人文特色与日常生活气息的风景。社区向人们展示了各种风格、形态的建筑,同时这些风格迥异的建筑,也带来了别具风情的立面感受。在立面设计中,采用了现代主义的设计风格,以简洁、实用、通透、少装饰为特点,而本项目则透过温暖诠释现代主义,释放它的能量,赋予建筑深情的人文关怀。

The project is located in the Xibao Mountain lot on the west side of Tian'e Plaza in Jingmen City. The base is featured by hilly and sloping lands. The whole terrain gradually slopes gently eastward and ends at the Fourth Trunk Canal on the east side.

The design seeks to make the buildings be a kind of landscape integrating cultural characteristics and the flavor of everyday life. The housing estate presents buildings of various styles, and meanwhile the characteristic buildings show unique facades. In facade design, modernistic design, which is featured by simplicity, functionality, transparency and less decoration, is adopted. The project interprets modernism with warmness and endows the buildings with deep humanistic care.

建设单位：荆门市凯凌房地产开发有限公司
功能用途：住宅
设计/竣工年份：2012年/～
建设地点：湖北荆门市
总建筑面积：451781 m²
总建筑高度/层数：56 m/18 F
结构形式：框架结构

Owner: Jingmen Kailing Real Estate Development Co., Ltd.
Function: Residence
Design/Completion Year: 2012/~
Construction Site: Jingmen City, Hubei
Total Floor Area: 451,781 m²
Total Height/Floor: 56 m/18 F
Structure: Frame Structure

贵阳市花溪区十和田项目
Guiyang Huaxi Shihetian Project

本项目位于贵阳花溪大道以西、贵州大学以北的10~13号四个地块,其中10、11、13号为住宅用地,12号为商业用地。项目总用地面积85 867 m²,用地总体西高东低,与东侧的湿地公园隔路相望,景观资源相当优越。

建筑立面采取现代简洁的风格,颜色为暖色调,与地块周围的自然山水环境形成适度的对比,建筑屋顶根据规划要求,采用现代坡屋顶的形式。

12号地块的公寓立面朝向花溪大道跌落,形成退台的屋顶花园,以生态建筑的形象为花溪大道打造了一道亮丽的风景。

The project is located in the 10-13 lots in the west of Huaxi Avenue and north of Guizhou University in Guiyang City, of which, Lot 10, 11 and 13 are residential lands while Lot 12 is commercial land. The project land, covering an area of 85,867 m², is higher in the west and lower in the east and faces the wetland park on the east side across the road, presenting superior landscape resources.
The building facade, in a modern and simple style and warm colors, forms a moderate contrast with the surrounding natural mountain and water. According to the planning, modern sloping roofs are adopted. The apartment facade in Lot 12, facing Huaxi Avenue and gradually dropping, forms terraced roof gardens, creating a beautiful view on Huaxi Avenue as ecological buildings.

结构形式：框筒结构

建 设 单 位：贵阳市花溪区十和田项目公司
功 能 用 途：住宅、单身公寓、商业
设计/竣工年份：2012年/~
建 设 地 点：贵州省贵阳市花溪区
总 建 筑 面 积：447 240 m²
总建筑高度/层数：100 m /33 F
结 构 形 式：框筒结构

Owner: Guiyang Huaxi Shihetian Project Corporation
Function: Residence, Bachelor Apartment, Commerce
Design/Completion Year: 2012/~
Construction Site: Huaxi District, Guiyang City, Guizhou
Total Floor Area: 447,240 m²
Total Height/Floor: 100 m /33 F
Structure: Frame-Tube Structure

黄岩羽村安置房工程
Huangyan Yucun Village Resettlement Housing Project

设计利用基地的主要形态，设置了一条南北向的空间序列轴线。住宅沿此轴线对称布置，形成一个有序而富有韵律感的空间。同时将小面宽的住宅设置在用地中部，整体交错排列，在获得足够的日照的同时，形成良好的景观视野。住宅的立面设计简洁、典雅，结合阳台和窗户的设计，形成活泼、有序的建筑表皮肌理，同时以竖向构架为立面主轮廓，强调建筑的挺拔感。加上窗户、栏板的玻璃与金属材料的引入以及石材和质感涂料的结合运用，使整个小区形成一种现代、大气、富有品质的建筑立面景观。

In design, based on the main terrain of the bot, a spatial sequence axis in the north-south direction is set, while the residential buildings are arranged symmetrically along the axis, forming an orderly and rhythmic spatial sequence. Meanwhile, the residential buildings with a small face width are arranged in the middle of the lot and are arranged in a staggered manner, so the buildings will have excellent landscape views when there's sufficient sunshine. The facade design of the residential buildings is simple and elegant. Combined with the balcony and window design, a lively and orderly architectural texture is formed. Meanwhile, the vertical framework as the main outline of the facade emphasizes the tallness and straightness of the building. With the introduction of the glass and metal materials of windows and boards as well as the combined usage of stone materials and quality paint, a modern, grand and high-quality facade view is formed.

建 设 单 位：台州市黄岩南区建设指挥部
功 能 用 途：住宅
设计/竣工年份：2011年/～
建 设 地 点：浙江省台州市黄岩区
总建筑面积：181 460 m²
总建筑高度/层数：57 m /18 F
结 构 形 式：框架结构

Owner: Huangyan South District Construction Office
Function: Residence
Design/Completion Year: 2011/~
Construction Site: Huangyan District, Taizhou City, Zhejiang
Total Floor Area: 181,460 m²
Total Height/Floor: 57 m/18 F
Structure: Frame Structure

江苏汇丰城市综合体方案
Jiangsu HSBC Urban Complex

建 设 单 位：金银岛国际大酒店有限公司
功 能 用 途：住宅、酒店、商业
设计/竣工年份：2012年/～
建 设 地 点：江苏省泰州市姜堰区
总 建 筑 面 积：370 694 m²
总建筑高度/层数：100 m/24 F
结 构 形 式：框架结构

Owner: Treasure Island International Hotel Co., Ltd.
Function: Residence, Hotel, Commerce
Design/Completion Year: 2012/~
Construction Site: Jiangyan District, Taizhou City, Jiangsu
Total Floor Area: 370,694 m²
Total Height/Floor: 100 m/24 F
Structure: Frame Structure

本项目位于江苏泰州市姜堰区，形态涵盖高层住宅、排屋、酒店、商业及辅助配套用房。

从地块西侧的改造河道引入水系，自然、流动而且纯净，从西往东，在高层住宅中心形成大型水景，从南向北，经过高层住宅区、联排别墅区，在酒店的南侧形成大型景观中心，最终汇聚于地块东北侧。同时，水系与车道、景观道相结合，将几个区块完整串联起来，并有效分隔。建筑形象定位为地中海风格，给人们以惬意感，同时一扫城市建筑的灰暗单调。

The project is located in Jiangyan District, Taizhou City, Jiangsu. It includes high-rise residential buildings, townhouses, hotels, commercial and subsidiary houses.
The water system is introduced from the reconstructed river channel on the west side of the lot, which is natural, flowing and clear running from west to east, and forms a large waterscape in the center of the high-rise buildings. Running from south to north, another water system runs across the high-rise building area and townhouse area and forms a large landscape center on the south side of the hotel. The two water systems finally converge on the northeast side of the lot. Meanwhile, the water systems, combined with the lanes and landscape paths, connect the several areas in series and also effectively separate them. The building image is positioned as Mediterranean style, which makes people happy and avoids the gloom and monotony of city buildings.

山东临沂"爱伦坡"
"Ailunpo" in Linyi, Shandong

建设单位：临沂健国房地产开发有限公司
功能用途：商业、住宅
设计/竣工年份：2011年/～
建设地点：山东省临沂市北城新区
总建筑面积：214 000 m²
总建筑高度/层数：100 m/32 F
结构形式：框剪结构

Owner: Linyi Jianguo Real Estate Development Co., Ltd.
Function: Commerce, Residence
Design/Completion Year: 2011/~
Construction Site: Beicheng New District, Linyi City, Shandong
Total Floor Area: 214,000 m²
Total Height/Floor: 100 m/32 F
Structure: Frame-Shear Wall Structure

项目位于山东临沂市北城新区，东至通达路，北至三和六街，南至沂蒙七路。

整个小区呈南低北高的走势，南侧为连排别墅区，北侧沿三和六街为高层住宅，天际线变化丰富。建筑采用A-Deco风格，以竖向线条为主，强调建筑物的高耸、挺拔，形成拔地而起、傲然屹立的非凡气势，表达出不断超越的人文精神和力量。通过新颖的造型、艳丽夺目的色彩以及豪华材料的运用，使建筑成为一种摩登艺术的符号。

The project, located in Beicheng New District, Linyi City, Shandong Province, connects Tongda Road in the east, Sanheliu Street in the north and Qimengqi Road in the south.
The whole residential district is lower in the south and higher in the north. On the south side are townhouses while on the north side are high-rise residential buildings along Sanheliu Street. The skyline is full of changes. In the A-Deco style, vertical lines are highlighted to emphasize the tallness and straightness of the buildings and showcase the spirit and strength of constantly transcending oneself. Through the fresh modeling, brilliant colors and luxury materials, the buildings become new symbols of modern art.

杭州寰宇天下
Hangzhou U World

项目位于位于杭州市滨江区，西兴大桥与复兴大桥之间，钱塘江畔。

项目分为D、E两块用地，其性质为住宅用地，由27~35层的高层住宅组成。

建筑造型力图体现出具有现代感和国际化的城市综合体形象。立面设计强调"帆"的寓意，利用简洁的弧线和轻盈通透的玻璃幕墙，化解原本厚重的建筑体量。"扬帆起航"的寓意使滨江的城市形象在这里得到进一步升华。

The project is located in Binjiang District, Hangzhou City, between Xixing Bridge and Fuxing Bridge and by the Qiantang River.

The project is divided into D and E Lot, which are both residential lands and are composed of 27~35-storey high-rise buildings.

It seeks to represent a modern and international urban complex image in the building appearance design. In facade design, the meaning of "sails" is emphasized, with simple arcs and light and transparent glass curtain walls to neutralize the original heavy building mass. The implied meaning of the facade—"setting sailing" brings the city image of Binjiang to a higher level.

建设单位：杭州世贸世盈房地产有限公司
功能用途：住宅
设计/竣工年份：2011年/2013年
建设地点：浙江省杭州市滨江区
总建筑面积：283 823 m²
总建筑高度/层数：116 m/35 F
结构形式：框剪结构
合作单位：香港华艺设计顾问（深圳）有限公司
　　　　　豪斯泰勒·张·思图德建筑设计咨询（上海）有限公司

Owner: Hangzhou Shimaoshiying Real Estate Co., Ltd.
Function: Residence
Design/Completion Year: 2011/2013
Construction Site: Binjiang District, Hangzhou City, Zhejiang
Total Floor Area: 283,823 m²
Total Height/Floor: 116 m/35 F
Structure: Frame-Shear Wall Structure
Cooperation Unit: Hong Kong Huayi Design Consultants(Shenzhen) LTD. /HZS Design(Shanghai)

杭长铁路等重点工程农居拆迁安置房
Resettlement Housing for Farmers for Major Projects like Hangzhou-Changsha Passenger-dedicated Line

项目位于杭州市萧山区新塘街道原吕才庄村，西面为市心南路，连接萧山市区。规划占地面积 153 536 m²，建设用地面积 134 858 m²，总投资约 187 360万元。

项目由11幢26层点式高层、8幢28层板式高层、1幢23层过渡安置房及沿街2~3层裙房组成。安置户数2 425 户。

The project is located in the original Lvcaizhuang Village, Xintang Subdistrict, Xiaoshan District, Hangzhou City. On the south side is South Shixin Road connecting the downtown of Xiaoshan District. The planning land area is 153,536 m², the construction land area is 134,858 m², and the total investment is about RMB1,873,600,000.

The project is composed of eleven 26-storey point-type high-rise buildings, eight 28-storey slab-type high-rise buildings, one 23-storey make-shift building and 2 or 3-storey annex buildings along the street. There are 2,425 relocated households.

建设单位：杭州市萧山区人民政府新塘街道办事处
功能用途：住宅
设计/竣工年份：2011年/～
建设地点：浙江省杭州市萧山区新塘街道
总建筑面积：446 000 m²
总建筑高度/层数：82 m /28 F
结构形式：框剪结构

Owner: Xiaoshan District People's Government Xintang Street Agency, Hangzhou
Function: Residence
Design/Completion Year: 2011/~
Construction Site: Xintang Street, Xiaoshan District, Hangzhou City, Zhejiang
Total Floor Area: 446,000 m²
Total Height/Floor: 82 m /28 F
Structure: Frame-Shear Wall Structure

三门银轮玫瑰湾小区
Sanmen Yinlun Rose Bay Residential District

建设单位：三门银轮置业发展有限公司
功能用途：住宅
设计/竣工年份：2011年/~
建设地点：浙江省台州市三门县
总建筑面积：60 405 m²
总建筑高度/层数：56 m/18 F
结构形式：框剪结构

Owner: Sanmen Yinlun Real Estate Development Co., Ltd.
Function: Residence
Design/Completion Year: 20011/~
Construction Site: Sanmen County, Taizhou City, Zhejiang
Total Floor Area: 60,405 m²
Total Height/Floor: 56 m/18 F
Structure: Frame-Shear Wall Structure

住宅追求质朴，摒弃烦琐的线脚和装饰，从图底关系上干净的体量也容易成为环境的背景。在轮廓上结合户型变化形成丰富的天际线，屋顶采用厚重线脚处理，让建筑更具整体感。阳台用薄板无框玻璃栏杆等造型元素营造光线和阴影，很多造型上均采用了独特的建筑语汇。颜色主要选用明快的暖色调，给居住者一种舒适、温馨的感觉。

It pursues simplicity in house design and abandons complicated architraves and decorations, as well as a clean mass which can easily become the environmental background. In outline design, it pursues rich skyline based on the changes in room layout, and the roofs are provided with heavy architraves, which together make the whole building look more integrated. The balconies are provided with thin-slab rimless glass railings to create light rays and shadow. The unique architectural languages are adopted. Most of the colors are warm and lively, to make residents feel warm and comfortable.

芜湖县南湖品园小区
Nanhu Pinyuan Residential District in Wuhu County

建设单位：芜湖玉湖置业有限公司
功能用途：住宅
设计/竣工年份：2011年/～
建设地点：安徽省芜湖市
总建筑面积：60 125 m²
总建筑高度/层数：51.35 m/18 F
结构形式：框剪结构

Owner: Wuhu Yuhu Real Estate Co., Ltd.
Function: Residence
Design/Completion Year: 2011/～
Construction Site: Wuhu City, Anhui
Total Floor Area: 60,125 m²
Total Height/Floor: 51.35 m/18 F
Structure: Frame-Shear Wall Structure

项目位于芜湖滨湖大道以北，罗福湖路以西。工程设计将地块划分为北侧高层公寓区、南侧低层联体住宅区以及与两个区域便捷相连的东侧配套会所区。北区共设计有4幢18层高层公寓，围合成1个住宅组团；南区设计有16幢3层联体住宅，形成相对独立的低层住宅区。单体设计中，户型力求明确的分区。功能用房集中布置，各居室面积分配合理，拥有良好的通风与日照条件以及优越的景观视野。造型上追求质朴、干净的风格，屋顶采用构架造型处理，让建筑更具整体感。

The project is located in the north of Binhu Avenue and south of Luofuhu Road in Wuhu. In the engineering design, the lot is divided into high-rise apartment zone in the north, low-rise townhouse zone in the south, and supporting club zone in the east that connects the two aforesaid zones for the seek of convenience. In the north zone, four 18-storey high-rise apartment buildings are designed, which form a residential cluster; in the south zone, sixteen 3-storey town houses are designed to form a relatively independent low-rise residential district. In the design of individual buildings, it seeks definite partitions in room layout. Functional houses are arranged in a centralized way, the area for each room is reasonably designed, and every room has excellent ventilation and sunshine condition, as well as superb landscape view. In appearance design, it pursues a simple and clean style. The roof with a framework design makes the building look more integrated.

杭州阳光郡
Hangzhou Sunny Estate

项目位于杭州市拱墅区祥符镇，地块东侧为城北进入市中心的主干道莫干山路。

小区设计充分利用地块本身及周边的现状，大部分住宅均按正南北向布置，两幢弧形板式住宅分别布置于南北两边，中间以单幢塔式住宅或两三个单元组成的板楼，围合出两个面积约为10 000 m²的开敞式超大中央花园。而在地块东西两边，分别布置两个单元的短板式住宅和点式塔楼，以丰富小区整体的空间感及提供良好的景观和自然的采光通风。整个小区的建筑基本上都南北错落布置，使得不同位置的单体建筑均有宽阔的景观视野、充足的日照以及良好的自然通风。

The project is located in Xiangfu Town, Gongshu District, Hangzhou City. On the east side of the lot is Moganshan Road, an arterial road in north of Hangzhou City leading to the downtown area.

The design of the residential district makes full use of the lot itself and the conditions of the surroundings. Most of the residential buildings are arranged facing the south. Two arc slab-type buildings are set on the south and north sides respectively, while between them are individual tower-type buildings or slab-type buildings consisting of 2 or 3 units, to form two open superlarge central gardens of about 10,000 m². While on the east and west sides of the lot, two-unit short-slab-type buildings and point-type tower buildings are arranged, to enrich the overall space of the residential district and provide excellent landscape views and natural lighting and ventilation. The buildings in whole are staggered, which provides open and wide landscape view, sufficient sunshine and excellent natural ventilation for individual buildings at different positions.

建 设 单 位：浙江建投发展房地产开发有限公司
功 能 用 途：住宅
设计/竣工年份：2011年/～
建 设 地 点：浙江省杭州市祥符镇
总 建 筑 面 积：350 000 m²
总建筑高度/层数：85 m /28 F
结 构 形 式：框剪结构

Owner: Zhejiang Jiantou Development Real Estate Development Co., Ltd.
Function: Residence
Design/Completion Year: 2011/~
Construction Site: Xiangfu Town, Hangzhou City, Zhejiang
Total Floor Area: 350,000 m²
Total Height/Floor: 85 m /28 F
Structure: Frame-Shear Wall Structure

萧山浪琴湾
Xiaoshan Langqin Bay

项目地处青六路和江东二路交会处。整个基地东西长约360 m，南北宽约205 m，总用地面积约68 986 m²，是包括住宅、商业以及一层地下室在内的综合建筑群。

设计通过本地块独特的地理位置和极具规模的建筑体量，形成江东工业园区重要的标志性核心建筑。通过功能的多样化和开阔的空间形态体现强烈的公共性与开放性，最大限度地包容城市生活，凝聚人气。在建筑表现手法上突出现代住宅所具有的现代化、信息化特点，建设成为以居住为中心的现代化居住区。

The project is located at the junction of Qingliu Road and 2nd Jiangdong Road. The whole base is about 360 m long in the east-west direction and about 205 m wide in the north-south direction, covering a total land area of about 68,986 m². It is a comprehensive building cluster including residential and commercial buildings and one-floor basement.

The design aims to create an important landmark building complex in the lot in Jiangdong Industrial Park by making use the unique geographical location of the lot and the large building mass, and to showcase its strong publicness and openness with diversified functions and an open and vast space to include urban living and gather people to the greatest extent. In the technique of architectural expression, the modernism and informatization of modern residences are highlighted. The project aims to be built into a modern residential area centered by residence.

A+B组合平面

A户型
建筑面积：89.1平方米
套内面积：63.0平方米

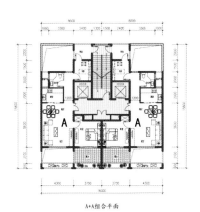

A+A组合平面

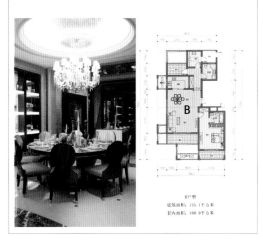

B户型
建筑面积：135.1平方米
套内面积：100.0平方米

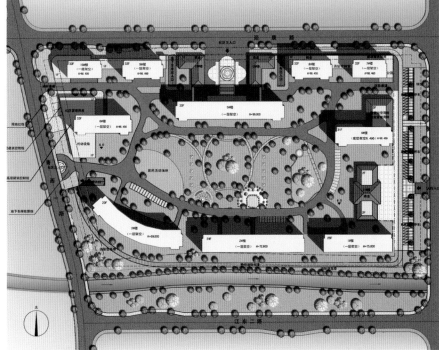

建 设 单 位：浙江铂丽置业有限公司
功 能 用 途：住宅
设计/竣工年份：2011年/～
建 设 地 点：浙江省杭州市萧山区江东工业园区
总 建 筑 面 积：261 657 m²
总建筑高度/层数：99.15 m /32 F
结 构 形 式：框剪结构

Owner: Zhejiang Boli Real Estate Co., Ltd.
Function: Residence
Design/Completion Year: 2011/~
Construction Site: Jiangdong Industrial Park, Xiaoshan District, Hangzhou City, Zhejiang
Total Floor Area: 261,657 m²
Total Height/Floor: 99.15 m /32 F
Structure: Frame-Shear Wall Structure

山东淄博"正承·PARK"
"Zhengcheng·PARK" in Zibo, Shandong

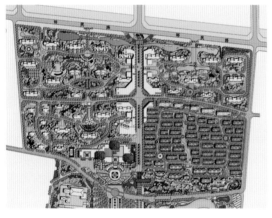

项目地处山东省淄博市北部,环境十分优美,景观资源极其丰富。该居住区划分为4个小区,居住区中部南北轴线方向为商业街区。居住区整体呈南低北高的走势,各个小区都有各自的空间系统和区域中心。

A区由洋房构成,中心引入水系,布置双拼别墅和联排别墅。B区由9幢高层构成,中心是大面积的小区绿化,整体空间布局大气。C区由14幢高层构成,空间布局在规律中求变化,中心景观轴线和大面积的公共绿地丰富了小区的空间形态。D区主要由18层的高层构成,鉴于交通、日照和空间高度的考虑,将酒店和幼儿园布置在D区。

The project, located in the north of Zibo City, Shandong Province, enjoys a beautiful environment and abundant landscape resources. The residential district is divided into four sub districts for overall planning. On the middle north-south axis of the residential district is a business block. The whole residential district is high in the north and low in the south, and each subdistrict has its own spatial system and regional center.

Subdistrict A is composed of western-style houses such as duplex villas and townhouses, with a central water system. Subdistrict B is composed of 9 high-rise buildings, with a large green area in the center. Subdistrict C is composed of 14 high-rise buildings. It seeks changes in regular pattern in spatial layout and the central landscape axis and large public green area enrich. the spatial pattern of the subdistrict. Subdistrict D is mainly composed of 18-storey buildings. Considering transportation, sun exposure and spatial height, hotels and kindergarten are set in Subdistrict D.

建设单位：淄博高新正承房地产开发有限公司
功能用途：商业、住宅
设计/竣工年份：2011年/～
建设地点：山东省淄博市高新区
总建筑面积：952 000 m²
总建筑高度/层数：100 m /33 F
结构形式：框剪结构

Owner: Gaoxin Zhengcheng Real Estate Development Co., Ltd.
Function: Commerce, Residence
Design/Completion Year: 2011/~
Construction Site: Gaoxin District, Zibo City, Shandong
Total Floor Area: 952,000 m²
Total Height/Floor: 100 m /33 F
Structure: Frame-Shear Wall Structure

恒隆国际花园及恒隆国际酒店
Henglong International Garden and Henglong International Hotel

建设单位：福建恒隆地产有限公司
功能用途：住宅、酒店、商业
设计/竣工年份：2010年/～
建设地点：福建省晋江安海镇梧埭村
总建筑面积：180 000 m²
总建筑高度/层数：100 m /33 F
结构形式：框剪结构

Owner: Fujian Henglong Real Estate Co., Ltd.
Function: Residence, Hotel, Commerce
Design/Completion Year: 2010/～
Construction Site: Wudai Village, Anhai Town, Jinjiang City, Fujian
Total Floor Area: 180,000 m²
Total Height/Floor: 100 m /33 F
Structure: Frame-Shear Wall Structure

项目位于安海镇梧埭村南环路东侧，是未来安海镇的发展方向，西临环镇公路，南面紧临乡镇河道。地块总用地面积 50 553 m²，分为三个区域，其中酒店用地面积 12 257.89 m²，住宅北区用地面积15 658.92 m²，住宅南区用地面积 226 36.7 m²。本工程用地现状为健身中心、游泳设施及物流场地等，地势平坦，南侧有自然河流通过，具有东南沿海城市的地质特色。本工程充分利用现有自然环境将地块营造成为城市高品质酒店及适宜居住的主题居住小区。

The project, located on the east side of South Ring Road in Wudai Village, Anhai Town , is the future development direction of Anhai Town. The project adjoins the ring highway in the west and the town river in the south. The lot, with a total land area of 50,553 m², is divided into three areas, of which, the land area used to build the hotel is 12,257.89 m², the north residential land area is 15,658.92 m², and the south residential land area is 22,636.70 m². There is currently a fitness center, swimming facilities and logistics sites. On the south side of the flat terrain, there is a natural river running by, showing the geological features of coastal cities in Southeast China. The project makes full use of the existing natural resources, trying to create a residential district with a high-quality hotel and a pleasant living environment.

萧山云涛名苑
Xiaoshan Yuntao Garden

建设单位：杭州易筑房地产开发有限公司
功能用途：住宅、商业
设计/竣工年份：2010年/2013年
建设地点：浙江省杭州市萧山闻堰镇
总建筑面积：120 000 m²
总建筑高度/层数：80 m /27 F
结构形式：框剪结构
合作单位：美国开朴建筑设计有限公司

Owner: Hangzhou Yizhu Real Estate Development Co., Ltd.
Function: Residence, Commerce
Design/Completion Year: 2010/2013
Construction Site: Wenyan Town, Xiaoshan District, Hangzhou City, Zhejiang
Total Floor Area: 120,000m²
Total Height/Floor: 80m/27F
Structure: Frame-Shear Wall Structure
Cooperation Unit: America CAPA Architectural Design Co., Ltd.

项目位于杭州市萧山区湘湖区块闻堰镇。

对住宅品质的追求要求赋予住宅完善的使用功能和审美功能，设计力争做到细节到位，可以满足各类功能性需求，同时创造出宜人的社区环境，包括自然环境、小区配套和人文环境。

本案的设计概念主题为"小空间，大庭院"。景观在总体规划的大格局下，顾及使用者对每种产品的每一处的感受，尤其注重近人尺度的庭院。小区庭院营造出中国传统式的"移步换景"之意境，再加以水景的配合，使得庭院更加具有灵性。

The project is located in Wenyan Town in Xianghu Lot, Xiaoshan District, Hangzhou City.
The pursuit of housing quality endows perfect useful functions and aesthetic function of residential houses. The design seeks to make every detail perfect to meet all functional requirements. Meanwhile, creating a pleasant community environment, including natural environment, supporting facilities and cultural environment .
The design concept of the project is "small space but large courtyard". Under the overall planning, designers take into consideration users' using experiences of every product, especially the courtyards which are closely related to privacy. The courtyards emphasize the traditional Chinese concept of "finding different views at every step". In combination with the waterscape, the courtyards are livelier.

山东郯城中央华庭
Central Garden in Tancheng, Shandong

项目用地位于山东省郯城县，用地面积139 998 m²，建筑主要为高层住宅公寓及配套公共建筑。

小区内共设计有37幢高层公寓，围合成4个住宅组团，景观内庭院与组团间的小区带状中心景观相得益彰。公寓为11层的板式高层住宅，户型设计以舒适性为主，强调均好性；造型设计简约，偏重经济性与实用性。小区配套商业用房布置在小区北侧及东侧沿街；物业用房布置在小区主入口广场的东西两侧；社区服务用房、幼儿园、医疗服务用房等布置在东侧沿街裙房。

The project is located in Tancheng County, Shandong Province with a land area of 139,998 m². The project is mainly composed of high-rise apartment buildings and supporting public buildings.
A total of 37 high-rise apartment buildings form four residential clusters. The inner courtyard landscape and the belt central landscape connecting the clusters can bring out the best in each other. The apartment buildings are 11-storey slab-type high-rise buildings. The design of the house layout emphasizes comfortable living and sharing of environmental resources, while the appearance design emphasizes simplicity, economy and practicability. The supporting commercial facilities of the residential district are arranged along the street in the north and east sides; the property management houses are set in the east and west sides of the main entrance square of the residential district; and the community service houses, kindergarten and medical service houses are set in the annex building along the street in the east side.

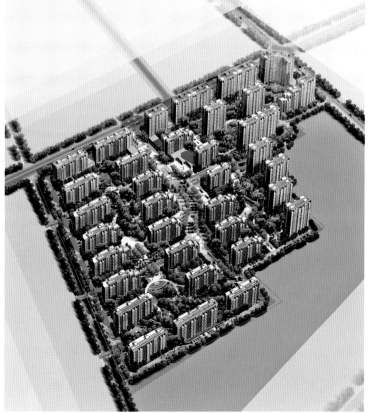

建设单位：临沂银海置业有限公司
功能用途：住宅
设计/竣工年份：2010年/2013年
建设地点：山东省临沂市郯城县
总建筑面积：383 105 m²
总建筑高度/层数：33.75 m /11 F
结构形式：框剪结构

Owner: Linyi Yinhai Real Estate Co., Ltd.
Function: Residence
Design/Completion Year: 2010/2013
Construction Site: Tancheng County, Linyi City, Shandong
Total Floor Area: 383,105 m²
Total Height/Floor: 33.75 m /11 F
Structure: Frame-Shear Wall Structure

当涂豪邦君悦华庭住宅小区
Dangtu Haobang Junyuehuating Residential District

建 设 单 位：当涂豪邦置业有限公司
功 能 用 途：住宅
设计/竣工年份：2009年/～
建 设 地 点：安徽省马鞍山市当涂县
总 建 筑 面 积：206 878 m²
总建筑高度/层数：94.85 m /3 F
结 构 形 式：框剪结构

Owner: Dangtu Haobang Real Estate Co., Ltd.
Function: Residence
Design/Completion Year: 2009/~
Construction Site: Dangtu County, Maanshan City, Anhui
Total Floor Area: 206,878 m²
Total Height/Floor: 94.85 m/32 F
Structure: Frame-Shear Wall Structure

小区依照"北高南低"的"风水学"格局灵活布置，在东北侧布置6幢32层住宅楼，东侧布置2幢17层住宅楼，西侧布置4幢18层住宅楼、1幢16层住宅楼、1幢17层住宅楼，通过高楼之间的围合形成中心大花园，同时使所有住户的主要房间均可以观赏优美的中心绿地景观，最大限度地挖掘了项目所在地块的环境资源，也充分利用了自身创造的内部景观，达到了资源的共享与共用。在小区的东北角布置一幢配套公建楼，内设农贸市场等与生活息息相关的配套措施，为住户提供极大的便利。

The residential district is flexibly arranged according to the layout of "higher in the north and lower in the south" based on the theory of Fengshui. On the northeast side of the residential district are six 32-storey residential buildings, on the east side are two 17-storey residential buildings, and on the west side are four 18-storey residential buildings, one 16-storey residential building and one 17-storey building. Based on the enclosure of the high-rise buildings, designers set a large central garden and make all residents' main rooms face the beautiful central green landscape. By doing so, it makes the best use of the environmental resources at the lot as well as the created internal landscape, which achieves resource sharing. At the northeast corner of the residential district is a public building which is equipped with supporting facilities closely related to everyday life such as farmers' markets, bringing great convenience to the residents.

浙江天台红石梁广场
Zhejiang Tiantai Hongshiliang Square

建设单位：浙江红石梁房地产开发有限公司
功能用途：商业、办公、住宅
设计/竣工年份：2009年/~
建设地点：浙江省台州市天台县
总建筑面积：154 079 m²
总建筑高度/层数：99.9 m/32 F
结构形式：框剪结构

Owner: Zhejiang Hongshiliang Real Estate Development Co., Ltd.
Function: Commerce, Office, Residence
Design/Completion Year: 2009/~
Construction site: Tiantai County, Taizhou City, Zhejiang
Total Floor Area: 154,079 m²
Total Height/Floor: 99.9 m/32 F
Structure: Frame-Shear Wall Structure

项目位于浙江省天台县，总规划建设用地面积约为32 600 m²。设计在面向赤城路位置以建筑裙房的形式，设置一个大型商业广场，充分利用了赤城路的商业价值。东侧接近规划用地线为排屋区。裙房北侧是高层住宅区，两幢弧形的高层住宅南北相错而立，围合出小区中心绿地空间。每个单元都有各自的入口空间和独立场地，互不干扰。高层住宅西侧靠近利民路为单身公寓，底层设有综合服务用房，物业管理用房位于北侧A楼架空层中。所有的住宅楼尽量按照南北朝向或南偏东布置。小区居民都可以享受日晒厅空间或眺景无限的美好生活。

This project is located in Tiantai Town, Zhejiang Province and the total designed area is 32600m². A large commercial square is designed to face Chicheng Road in the form of building podium, which makes the best use of the commercial value of Chicheng Road. The east area is designed to be the terrace area. To the north side of the podium is high-rise residential area. Two cambered high-rise buildings are located in the south and in the north, which encloses a green area for the center of the community. Each unit has independent entrance and area and does not bother each other. The west side of the high-rise building close to Limin Road is the apartments for singles. There is comprehensive service building in the ground floor. The property management building is in the overhead layer in Building A in the north. All the residential buildings face towards the south and north or south by east. All the residents in the community can enjoy a balcony with sunshine or the infinite scenery for better life.

体育建筑
Sports Building

264	临海市体育文化中心方案
266	浙江工商大学文体中心
268	奉化体育馆
269	杭州市全民健身中心方案
270	岱山县文化体育中心
272	湖州南太湖湿地奥体公园

264	Design Scheme for Linhai Sports Culture Center
266	Cultural and Sports Center in Zhejiang Gongshang University
268	Fenghua Gymnasium
269	Design Scheme for Hangzhou Citizen Fitness Center
270	Daishan County Cultural and Sports Center
272	Huzhou South Taihu Wetland Olympic Park

临海市体育文化中心方案
Design Scheme for Linhai Sports Culture Center

项目位于临海市，作为一座文化综合体，整合了博物馆、规划馆、文化馆、剧院、图书馆、档案馆六大功能。设计以抽象的现代形体表现百舸争流、迎风破浪的气势，突出建筑的文化特性。建筑主体以两大一小的梭形船作为组合元素，其中最大的体量位于用地的北侧居中，其尾部与小的船形组合成一个有机整体，内设剧院、文化艺术中心、档案馆、图书馆等。另外一个区域为博物馆和规划馆。三者之间形成三角的呼应关系。设计采用地景建筑的设计模式，将广场、平台、屋面的界限模糊，强调建筑使用者的行为设计，同时结合灵动活泼的平台设计，不仅丰富了场地景观环境，更有序地组织和引导了使用者的空间行为动向。

This project is located in Linhai City. As a cultural complex, it provides six main functions, including museum, planning hall, cultural hall, theater, library and archives. This design proposal uses modern abstract appearance to express its strong vigor and to highlights the cultural features of the building. The building consists of three fusiform volumes, two of which are big and one is small. The biggest volume is positioned in the middle of north border on the site and its end cooperates with the small volume to form an organic body, which contains theater, cultural and arts center, archives and library, etc. Museum and planning hall are arranged in the other volume. These three volumes form a triangular layout. This proposal uses landform building mode to blur the boundary of square, platform and roof and to highlight users' behavior design. The vivid platform design not only enriches the landscape environment on the site, but also effectively organizes and guides users' space movement direction.

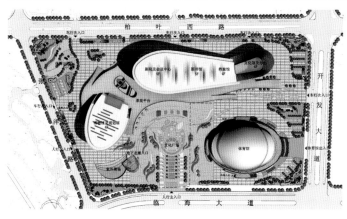

建 设 单 位：临海市建设局
功 能 用 途：博物馆、规划馆、文化馆、剧院、图书馆、档案馆
设计/竣工年份：2012年/～
建 设 地 点：浙江省临海市
总 建 筑 面 积：175 000 m²
总建筑高度/层数：37.4 m /5 F
结 构 形 式：框架、网架结构

Owner: Linhai Municipal Construction Bureau
Function: Museum, Planning Hall, Cultural Hall, Theater, Library, Archives
Design/Completion Year: 2012/~
Construction Site: Linhai City, Zhejiang
Total Floor Area: 175,000 m²
Total Height/Floor: 37.4 m /5 F
Structure: Frame-Structure, Grid Structure

浙江工商大学文体中心
Cultural and Sports Center in Zhejiang Gongshang University

项目位于浙江工商大学下沙校区内西北角，建设用地面积约为 23 000 m²。

本工程体育建筑等级为乙级，按座席规模属中型，共设座席 5 415 个，可进行篮球、排球、手球、网球、乒乓球、体操等比赛及大型室内文艺表演、集会。设计力求在满足各种体育比赛功能的基础上，结合实际需求和面向社会开放的要求进行设计。

建筑形体上取意"太阳花"，是对运动、波浪元素、弧形元素的取向，体现了一种诗意的浪漫，仿生手法的运用表达了人类对生命的尊重、对体育运动的热爱，体现了一种积极向上的生活态度。

The project is located at northwest corner in Xiasha campus of Zhejiang Gongshang University, and it covers a site area of about 23,000 m².

This is a Grade-B sports building having 5,415 seats and it could be used to host basketball, volleyball, handball, tennis, pingpong and gymnastics matches, as well as large indoor performance and conference. The design proposal tries to provide functions for various sports games and to satisfy practical demands and social requirements.

The building has a sunflower shape composed of sports, wave and arc elements, to embody a poetic romance, to show human's respect to life and sports games, and to express a kind of positive living attitude.

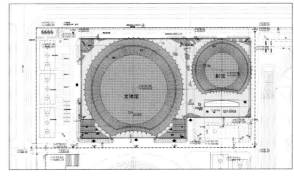

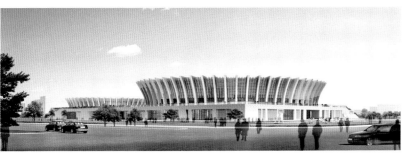

建 设 单 位：浙江工商大学
功 能 用 途：体育、演出
设计/竣工年份：2011年/～
建 设 地 点：浙江工商大学下沙校区
总 建 筑 面 积：19 149 m²
总建筑高度/层数：25 m /3 F
结 构 形 式：框架、钢网壳结构

Owner: Zhejiang Gongshang University
Function: Sports, Performance
Design/Completion Year: 2011/~
Construction Site: Xiasha Campus of Zhejiang Gongshang University
Total Floor Area: 19,149 m²
Total Height/Floor: 25 m /3 F
Structure: Frame-Structure, Steel Grid Shell Structure

奉化体育馆
Fenghua Gymnasium

建设单位：奉化市文体局
功能用途：体育
设计/竣工年份：2011年/～
建设地点：浙江省奉化市
总建筑面积：28 460 m²
总建筑高度：31.5 m
结构形式：框桁架结构

Owner: Fenghua Municipal Cultural and Sports Bureau
Function: Sports
Design/Completion Year: 2011/~
Construction Site: Fenghua City, Zhejiang
Total Floor Area: 28,460 m²
Total Height: 31.5 m
Structure: Frame-Truss Structure

项目位于拟建行政中心区块内，建筑主体呈南北向布置，在和大城路及桥东岸路两条主要城市道路之间均留有较大景观空间。体育馆比赛场地设于场馆南侧，面向南侧主入口，训练馆设于场地北侧。建筑形态上，体育馆选用寓意为船桨的竖向线条作为立面的主要肌理，各个立面均呈现出不同的富有韵律、逐步升高的弧度，形成柔美的曲线，恰似一浪推一浪，既体现了运动员拼搏奋斗的精神，又暗喻城市发展水平的稳步提高。在平面布局上，科学地安排各项功能，注重建筑布局给运动员与观众带来使用上的便捷性与空间感受上的舒适性。

The project is located within the administrative center plot which is under planning. The main building faces to south and north, reserving wide viewing space between Dacheng Road and Qiaodongan Road, which are two urban arteries. The courts are arranged at southern part of the gymnasium, facing to the south main entrance, while the training court is designed at the northern part. in the tenns of building shape, the gymnasium adopts vertical strips to create a quant-like appearance, and different bending degrees produce abundant increasing rhythms on each facade and form elegant curve like rushing waves, not only embodying athletes' struggling spirit, but also symbolizing the stable growth of urban development. As for the plan layout, various functions are reasonably and scientifically arranged to provide convenience and comfortable space experience for athletes and audiences.

杭州市全民健身中心方案
Design Scheme for Hangzhou Citizen Fitness Center

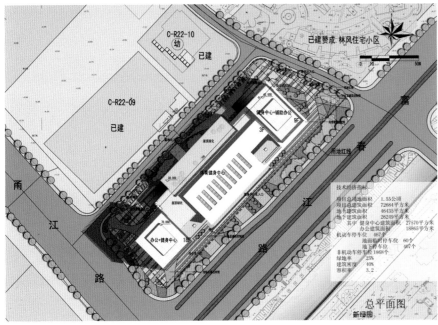

建设单位：杭州市体育发展集团
功能用途：健身中心
设计/竣工年份：2011年/～
建设地点：浙江省杭州市上城区
总建筑面积：72 684 m²
总建筑高度/层数：70 m/18 F
结构形式：框架结构

Owner: Hangzhou Sports Development Group
Function: Fitness Center
Design/Completion Year: 2011/~
Construction Site: Shangcheng District, Hangzhou City, Zhejiang
Total Floor Area: 72,684 m²
Total Height/Floor: 70 m/18 F
Structure: Frame Structure

项目用地位于杭州市上城区南星单元东部。整个地块大致呈长方形。设计围绕"开放""综合""简约""舒适"四大主题来塑造全民健身中心的建筑空间形态。项目采用集中式布局，尽量退让出广场、绿地。在用地的最西南端布置了一幢18层的高层建筑，其东侧为9层的管理办公用房，二者通过大体量的综合体育健身馆相接。

主楼外立面采用幕墙，体现了公共建筑谦和、沉静、稳健、简洁的特性。裙房采用大块的石材，通过运动馆自身的大体量来组合形体，使建筑更有冲击力，同时方正的体形也体现了建筑的严肃感。

The project is located on an approximately rectangular site at eastern part of Nanxing Block in Shangcheng District, Hangzhou City. The design scheme focuses on four themes, including "open", "complex", "simple" and "comfort", to create the spatial shape of the citizen fitness center. Centralized layout is adopted to save areas for square and green land as large as possible. An 18-storey high-rise building stands at the southwest end of the site and faces to a 9-storey office building at east, and they are connected with each other through large sports fitness complex.
Curtain wall is used on the main building facade to show the gentle, sober, stable and simple feature of public building. The podium building is decorated with massive stone, its huge and square volume produces stronger visual impact and expresses the solemnity of the building.

岱山县文化体育中心
Daishan County Cultural and Sports Center

项目位于岱山竹屿新区东部沿海地块，由椭圆形的主馆及水滴形训练馆组合而成，主馆屋顶和训练馆屋顶在空间形态上一气呵成，使建筑宛如海里跳跃的浪花，充满动感和韵律。该体育馆既有现代体育建筑的活力，又能体现沿海城市的建筑特色，与周围沿海景观相得益彰。体育中心按乙级体育建筑标准设计，部分指标达到甲级体育建筑标准，设计总座位数为3 072个，其中2 358个为固定座席，其余为活动座席。看台以南北向布置为主，属中型体育建筑。馆内设比赛场地，可举办手球、篮球、排球、乒乓球等比赛，同时亦可举办演唱会等文艺演出。

The project is located on a coastal plot at eastern area of Zhuyu New District, Daishan City. The design proposal consists of elliptic main hall and water-drop-shaped training hall. Their roofs realize a spatial continuity, infusing dynamic rhythm into the building which looks like rolling wave on the sea. This gymnasium not only embodies the vigor of modern sports building, but also expresses building features in coastal city, successfully integrating itself into the contextual coastal landscape. The sports center is designed according to standards for Grade-B sports building and some indexes have met to standards for Grade-A sports building. It provides 3,072 seats in total, among which 2358 seats are fixed and the rest are moveable. The stand stretches mainly along north-south direction. This project is a medium-scale sports building. It could be used to host sports games, such as handball, basketball, volleyball and pingpong, etc., and even concert.

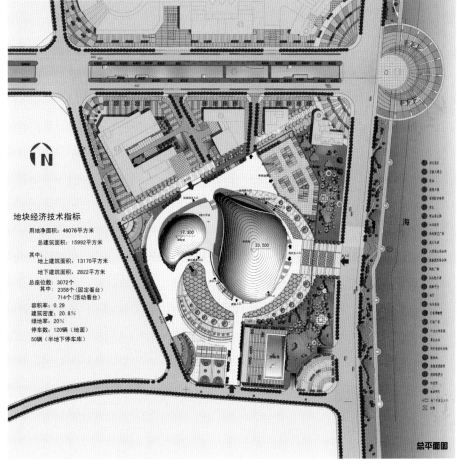

建设单位：岱山县海蓝城市建设有限公司
功能用途：体育
设计/竣工年份：2010年/2013年
建设地点：浙江省岱山市竹屿新区
总建筑面积：15 000 m²
总建筑高度：33 m
结构形式：框架、网架结构

Owner: Daishan Hailan Urban Construction Co., Ltd.
Function: Sports
Design/Completion Year: 2010/2013
Construction Site: Zhuyu New District, Daishan City, Zhejiang
Total Floor Area: 15,000 m²
Total Height: 33 m
Structure: Frame Structure, Grid Structure

湖州南太湖湿地奥体公园
Huzhou South Taihu Wetland Olympic Park

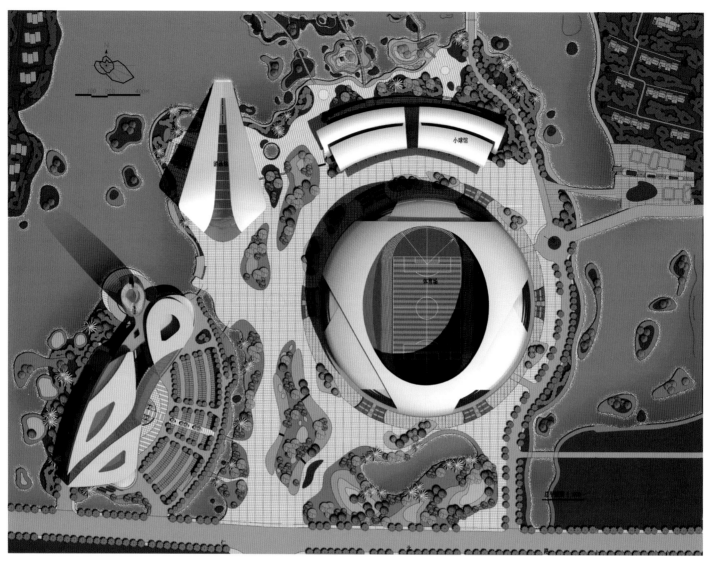

项目位于仁皇山新区奥体湿地公园中的核心区块，总用地面积31.3 hm²。工程包括体育场、游泳馆和小球中心。其中体育场按满足乙级体育赛事标准、40 000座规模建设；游泳馆按满足乙级体育赛事标准、观众席位1 500座规模建设；小球中心设羽毛球、乒乓球、壁球等球场及健身房。主体育场平面采用"内椭圆外圆"的布置方式。小球中心平面采用向心圆的布置方式。建筑屋盖为一个扇形平面，围绕背后的主体育场进行放射状布置。建筑由两个拱形的形体的排列构成，犹如层层波浪冲刷在平缓的湿地湖岸。

This project is located on a central plot of Olympic Wetland Park in Renhuangshan New District, covering a site area of 31.3 hm². It consists of stadium, swimming pool and small-ball center. The stadium has 40,000 seats for hosting Grade-B sports games; the swimming pool has 1,500 seats for hosting Grade-B sports games; small-ball center contains badminton court, pingpong court, squash court and fitness center. The main stadium adopts an "inner oval and outer annular" plan layout, while the small-ball center is planned in a centripetal layout. The fan-shaped roof radiates around the back main stadium. It is composed of two arched elements which look like waves rushing on the wetland lakeshore.

建 设 单 位：湖州中兴建设开发公司
功 能 用 途：体育
设计/竣工年份：2010年/~
建 设 地 点：浙江省湖州市南太湖
总建筑面积：116 000 m²
总建筑高度：50 m
结 构 形 式：框架、钢网壳结构
合 作 单 位：德杰盟工程技术有限公司

Owner: Huzhou Zhongxing Construction and Development Company
Function: Sports
Design/Completion Year: 2010/~
Construction Site: South Taihu, Huzhou City, Zhejiang
Total Floor Area: 116,000 m²
Total Height: 50 m
Structure: Frame Structure, Steel Grid Shell Structure
Cooperation Unit: J.S.K SIAT Engineering Technique Co., Ltd.

规划与城市设计
Plan and City Design

276 杭州未来科技城"城市绿心"规划
278 杭州良渚组团中央商务区城市设计
280 杭州未来科技城西溪科技岛城市设计
282 湄洲湾职业技术学院迁建工程修建性详细规划
284 嘉兴子城片区城市有机更新概念规划与城市设计
286 杭州瓶窑老镇区概念规划
288 嘉兴湖滨片区城市有机更新概念规划与城市设计
290 杭州未来科技城重点建设区域城市色彩规划
292 栖霞滨湖新城控制性详细规划及重点地段城市设计
294 诸暨市枫桥镇枫江文化体育公园景观设计

276 "Urban Green Center" Planning in Hangzhou Future Sci-Tech City
278 Urban Design of Central Business District in Liangzhu, Hangzhou
280 Urban Design of Xixi Sci-Tech Island in Hangzhou Future Sci-Tech City
282 Construction Planning Details about Relocation of Meizhouwan Vocational Technology College
284 Organic Renovation Conceptual Planning and Urban Design of Jiaxing Zicheng Plot
286 Conceptual Planning of Pingyao Old Township of Hangzhou
288 Urban Organic Renovation Conceptual Planning and Urban Design of Waterfront Plot in Jiaxing City
290 Urban Color Planning at Key Construction Area of Hangzhou Future Sci-Tech City Project
292 Controlled Planning Details and Urban Design of Important Locations of Qixia Waterfront New City Project
294 Landscape Design of Fengqiao Town Fengjiang Cultural and Sports Park, Zhuji

杭州未来科技城"城市绿心"规划
"Urban Green Center" Planning in Hangzhou Future Sci-Tech City

本案位于未来科技城南端,坐拥五常、闲林两片湿地,东依绕城高速,南接02省道,西靠东西大道,北临文一西路。规划以"城·无界"为核心理念,在"营建多元复合的城市公共服务核心"和"绿色休闲的生态核心"两条线索的指引下,创造集商业、办公、文化、教育、休闲、旅游、生态等多元复合功能于一体的城市绿心。规划形成"一心两翼辉映,双轴多区联动"的总体发展结构,通过"跨界整合、消界融合、破界缝合"的理念,构建"无界共生、玉带环腰、湿地青袭"的核心特点,以开发促保护,最终打造一个充满活力的城市绿心,一个湿地与城市相交融的无界之城。

The project is located at southern border of the Future Sci-Tech City and is surrounded by Wuchang Wetland and Xianlin Wetland. It is adjacent to the ring expressway at east, 02 provincial road at south and Dongxi Boulevard at west and Wenyi West Road at north. The planning is based on the concept of "city • boundless" and focuses on two clues including "pluralistic and compound urban public service center" and "green ecological leisure center", to create an urban green center providing various functions, such as commerce, office, culture, education, leisure, tourism and ecological environment, etc. This results a general development structure of "one center cooperating with two wings, dual axes linking with multiple districts" and creates the core features of "boundless coexistence, prosperous central zone and wetland environment", through the concept of "cross-border integration, boundless cooperation, border breaking". This protection-oriented development successfully produces a vigorous urban green center and a boundless city where wetland and urban space intertwine with each other.

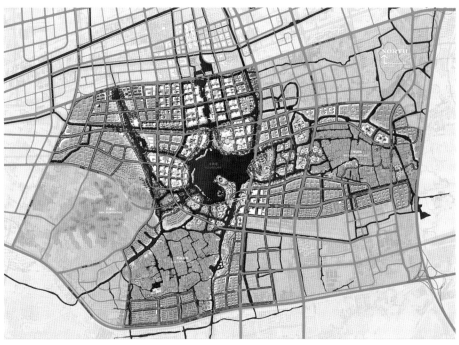

建设单位：杭州未来科技城管理委员会	Owner: Hangzhou Future Sci-Tech City Management Committee
功能用途：科技新城	Function: New Sci-Tech City
设计年份：2013年	Design Year: 2013
建设地点：浙江省杭州未来科技城	Construction Site: Hangzhou Future Sci-Tech City, Zhejiang
总建筑面积：11 000 000 m²	Total Floor Area: 11,000,000 m²
规划用地：18.8 km²	Planning Land: 18.8 km²

杭州良渚组团中央商务区城市设计
Urban Design of Central Business District in Liangzhu, Hangzhou

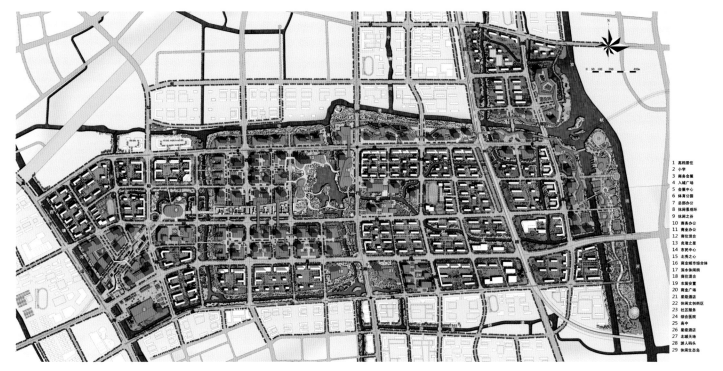

本案位于良渚组团南侧，紧临主城拱墅的勾庄区块，规划范围为5.2 km²。本案通过战略高度的把握、独特视角的切入和整体严谨的分析，提出了明确的发展目标、清晰的空间结构与合理的开发时序，最终将良渚组团中央商务区打造成为以商贸商务为主导，以综合服务和现代物流服务为支撑，兼具发展总部经济、金融商贸、旅游观光、行政办公、高新研发、文化娱乐及房地产业等多元功能，是彰显良渚新品质，塑造城北新地标的绿色高效之城。

This project is located in southern area of Liangzhu and is adjacent to Gouzhuang plot in Gongshu District, covering a planning area of 5.2 km². After effective control on strategic purpose, precise analysis on unique viewpoint, this scheme defines clear development objects, legible space structure and reasonable development sequence. It will build the central business district of Liangzhu into a green and effective district integrating various functions, such as commercial trade, business office, general service, modern logistics service, headquarters economy, financial and commercial trade, tourism, administrative office, high-tech research and development, cultural entertainment and real estate development, etc.

建设单位：杭州良渚组团管理委员会
功能用途：城市副中心
设计年份：2013年
建设地点：浙江省杭州市良渚
总建筑面积：7 240 000 m²
规划用地：520 hm²

Owner: Hangzhou Liangzhu Management Committee
Function: Urban Subcenter
Design Year: 2013
Construction Site: Liangzhu, Hangzhou City, Zhejiang
Total Floor Area: 7,240,000 m²
Planning Land: 520 hm²

杭州未来科技城西溪科技岛城市设计
Urban Design of Xixi Sci-Tech Island in Hangzhou Future Sci-Tech City

本案位于杭州未来科技城东南部，坐拥五常湿地，东依绕城高速，南接常余路，西靠闲林港，北临文二西路。

本案以"隐·逸"为核心理念，在"营建和谐生态的湿地氛围"和"激发多元复合的城市活力"两条线索的指引下，形成"无界共生——城市与湿地的交融、文脉延续——场所记忆的扬弃、珠联水脉——横向主轴的延展、涟漪渐进——圈层结构的布局"等核心特点。

规划通过"理水营脉—开湖筑岛—拥水塑心—圈层展开"，逐步推演，最终形成"一脉三心、绿廊通达、多区联动"的规划结构，打造一个充满活力、生态平衡、山水共栖的隐逸之城。

This project is located at the southeastern portion of Hangzhou Future Sci-Tech City and is surrounded by Wuchang Wetland. It is near to the ring expressway at east, connected to Changyu Road at south, adjacent to Xianlin Harbor at west and West Wener Road at north.

This scheme is designed on the core concept of "concealment and comfort" and guided by two clues including "building harmonious ecological wetland environment" and "activating pluralistic and compound urban vigor", to form such core features as "boundless coexistence for integration between urban area and wetland; contextual extension for development and discard of location memory; waterfront landscape for transverse axial expansion; waving ripple for layered annular layout, etc."

Through "water treatment—building island on lake—creating central elements—layered annular layout", this planning finally results a structure of "one axis and three cores, green corridor, connected areas" and creates a concealed and comfortable city full of vigor and ecological balance, where mountainous landscape and waterscape coexist.

建设单位：杭州未来科技城管理委员会
功能用途：高端总部、居住
设计年份：2013年
建设地点：浙江省杭州未来科技城
总建筑面积：2 000 000 m²
规划用地：5.8 km²

Owner: Hangzhou Future Sci-Tech City Management Committee
Function: High-end Headquarters, Residence
Design Year: 2013
Construction Site: Hangzhou Future Sci-Tech City, Zhejiang
Total Floor Area: 2,000,000 m²
Planning Land: 5.8 km²

湄洲湾职业技术学院迁建工程修建性详细规划
Construction Planning Details about Relocation of Meizhouwan Vocational Technology College

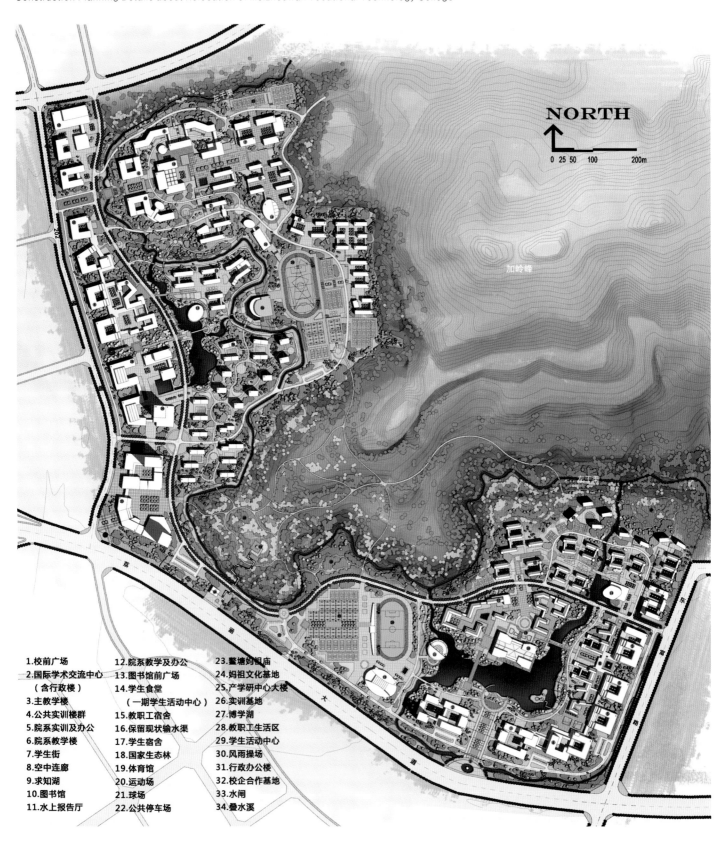

1. 校前广场
2. 国际学术交流中心（含行政楼）
3. 主教学楼
4. 公共实训楼群
5. 院系实训及办公
6. 院系教学楼
7. 学生街
8. 空中连廊
9. 求知湖
10. 图书馆
11. 水上报告厅
12. 院系教学及办公
13. 图书馆前广场
14. 学生食堂（一期学生活动中心）
15. 教职工宿舍
16. 保留现状输水渠
17. 学生宿舍
18. 国家生态林
19. 体育馆
20. 运动场
21. 球场
22. 公共停车场
23. 鳌塘妈祖庙
24. 妈祖文化基地
25. 产学研中心大楼
26. 实训基地
27. 博学湖
28. 教职工生活区
29. 学生活动中心
30. 风雨操场
31. 行政办公楼
32. 校企合作基地
33. 水闸
34. 叠水溪

本案位于莆田市涵江区，是莆田大学新区的重要组成部分。设计在"自然的大学——营建以学生为本的山水生态校园"和"共享的大学——打造与城市共融的人文智慧综合体"两条线索的指引下，力求打造一个以"山水·院街"为核心理念，以"生态校园、活力校园、共享校园、文化校园"为指导，传承莆田文脉，融合自然山水的现代化综合性校园。

This project, located in Hanjiang District of Putian City, is an important portion of Putian University New District. The design is based on two clues including "natural university— to create student-oriented ecological landscape campus" and "open university— to create a humanism wisdom complex in urban space", to create a modern compound campus focusing on the core concept of "landscape·courtyard street" and the vision of "ecological campus, vigorous campus, open campus and cultural campus", inheriting Putian history and integrating into the natural landscape.

建 设 单 位：湄洲湾职业技术学院
功 能 用 途：大专院校
设 计 年 份：2013年
建 设 地 点：福建省莆田市
总建筑面积：942 000 m²
规 划 用 地：137.6 hm²

Owner: Meizhouwan Vocational Technology College
Function: College
Design Year: 2013
Construction Site: Putian City, Fujian
Total Floor Area: 942,000 m²
Planning Land: 137.6 hm²

嘉兴子城片区城市有机更新概念规划与城市设计
Organic Renovation Conceptual Planning and Urban Design of Jiaxing Zicheng Plot

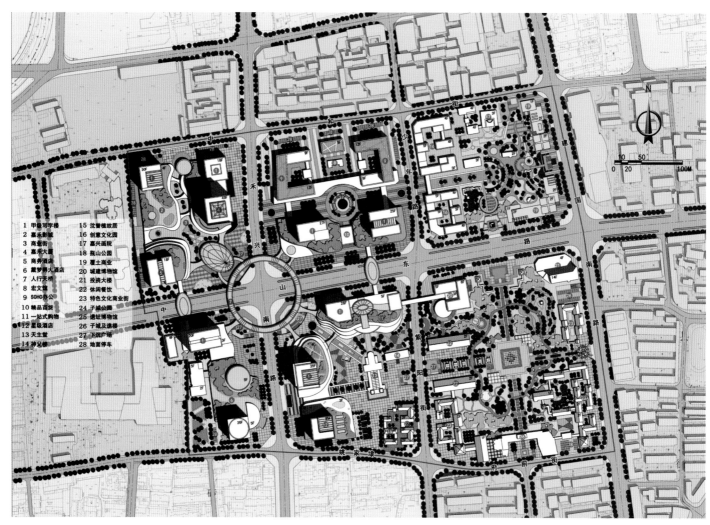

规划区地处嘉兴老城心脏地带，历史文化优势资源集聚。方案取"嘉禾兴城"之意，将规划区打造成彰显地域特色的市级商业商务中心与标志性公共活动场所。

本案通过构建商业集聚核、构建活力功能环形成"一环穿城，两轴联动，三心辉映"的发展构架；规划强调高度开发的城市综合体与历史遗存的有机融合，运用创造性手法演绎传统文化，不仅满足现代服务需求，而且实现了文脉传承；另外，通过人行天桥、建筑连廊、下沉广场构建立体便捷的步行体系，将商业购物、商务办公与休闲游憩等整合为一体，创造可体验性场所。

The planning zone is located in the old central area of Jiaxing City and enjoys abundant historical and cultural resources. This scheme takes the meaning of "economy-oriented urban prosperity" to build the planning area into a municipal-level commercial and business center and public landmark location boasting of regional features.

This scheme creates commercial center and vigorous ring to form a development structure of "one ring passing through the urban area, interlocking with two axes and cooperating with three central zones"; the planning emphasizes organic combination between modern urban complex and historical relic, uses creative manners to express traditional culture, not only satisfying modern service requirements, but also realizing historical inheritance; in addition, shopping mall, offices and leisure space are integrated to create an experience location through convenient stereo pedestrian system composed of pedestrian bridge, building corridor and sunk square.

建设单位：嘉兴市城乡规划建设管理委员会
功能用途：商业、商务、文化
设计年份：2013年
建设地点：浙江省嘉兴市
总建筑面积：655 600 m²
规划用地：31.4 hm²

Owner: Jiaxing Country and Urban Planning Construction Management Committee
Function: Commerce, Business and Culture
Design Year: 2013
Construction Site: Jiaxing City, Zhejiang
Total Floor Area: 655,600 m²
Planning Land: 31.4 hm²

杭州瓶窑老镇区概念规划
Conceptual Planning of Pingyao Old Township of Hangzhou

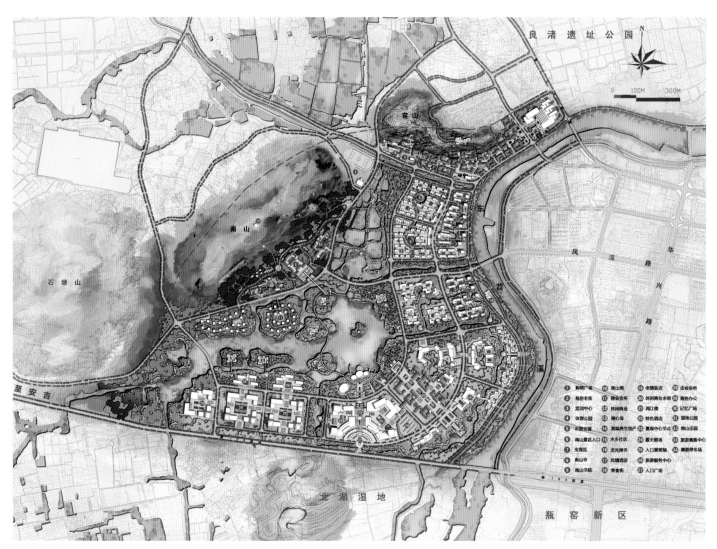

项目位于瓶窑镇西部、东苕溪西岸，北依良渚遗址，南连北湖湿地，内拥南山和下洋湿地景观，南北各有新老104国道通过，旅游资源丰富，对外交通便捷，总规划范围为2.84 km²。

规划充分研究了老镇区在瓶窑组团和"大径山"旅游体系中的角色与地位，依托其区位优势，深入发掘南山、下洋湿地自然资源和佛道、陶窑文化资源，以"悠游南山、乐享西岸"为规划理念，将老镇区打造成集旅游集散、生态人居、主题游乐、休闲购物、商务会议、民俗观光等多功能于一体的风情小镇。

The project is located in western area of Pingyao Town, west bank of Dongshao River, and it is adjacent to Liangzhu relic at north and Beihu Wetland at south, possessing Nanshan and Xiayang Wetland, and having abundant tourism resources and convenient traffic conditions with new and old 104 national highways passing through the site at south and north respectively. Total planning area of this project is 2.84 km².

The planning gives sufficient study on the role and importance of old township in Pingyao and "Dajing Mountain" tourism system, and depends on its geological advantages to explore the natural resources, such as Nanshan and Xiayang Wetland, etc., and cultural resources, such as Buddhism, Daoism and pottery kiln, etc. This scheme is based on the planning concept of "romantic Nanshan tourism and beautiful west bank" to build the old township into a resort integrating various functions, such as tourism, ecological and habitable environment, thematic entertainment, leisure, shopping, business conference and historical sightseeing, etc.

建 设 单 位：杭州市余杭区瓶窑镇人民政府
功 能 用 途：居住、商业、旅游
设 计 年 份：2012年
建 设 地 点：浙江省杭州市余杭区瓶窑镇
总建筑面积：930 000 m²
规 划 用 地：284 hm²
合 作 单 位：杭州市规划局余杭规划分局

Owner: Pingyao Town People's Government
of Yuhang District, Hangzhou
Function: Residence, Commerce, Tourism
Design Year: 2012
Construction Site: Yuhang District,
Hangzhou City, Zhejiang
Total Floor Area: 930,000 m²
Planning Land: 284 hm²
Cooperation Unit: Yuhang Planning Branch,
Hangzhou Planning Bureau

嘉兴湖滨片区城市有机更新概念规划与城市设计
Urban Organic Renovation Conceptual Planning and Urban Design of Waterfront Plot in Jiaxing City

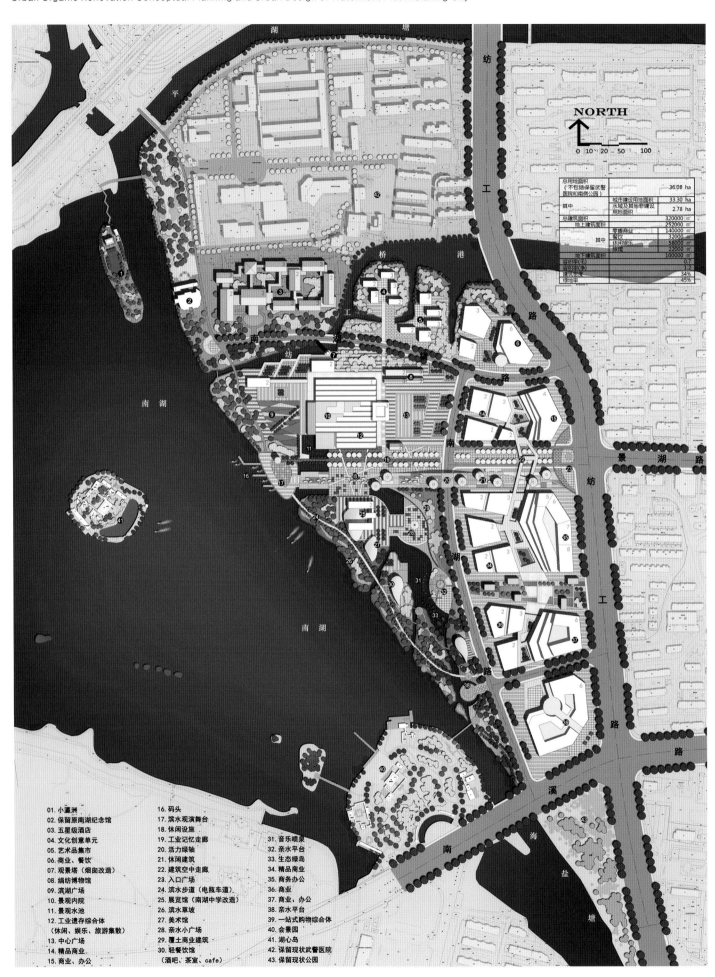

建 设 单 位：嘉兴市城乡规划建设管理委员会
功 能 用 途：滨湖商业及旅游服务中心
设 计 年 份：2013年
建 设 地 点：浙江省嘉兴市
总建筑面积：352 000 m²
规 划 用 地：36 hm²

Owner: Jiaxing Country and Urban Planning Construction Management Committee
Function: Waterfront Business and Travel Service Center
Design Year: 2013
Construction Site: Jiaxing City, Zhejiang
Total Floor Area: 352,000 m²
Planning Land: 36 hm²

　　项目位于嘉兴南湖东岸，与行政文化中心和老城商业金融核心隔南湖相望。方案以打造"城湖绿链"为规划理念，以多元复合的城市滨水公共活动中心和现代滨湖旅游服务中心的营建为线索，将湖滨区块打造成为以滨水活动为特色，以休闲娱乐和旅游服务为主导，以商业商务办公为支撑，兼具发展文化创意、艺术观演、博览展示等功能的南湖活力中心。规划以"一心引领、横轴连接、圈层环抱"作为总体布局结构，构建"水带环绕、绿廊串珠、蓝网绿脉"的景观格局，并形成"F"形的主干路网系统作为交通支撑。

The project is located by the east bank of Nanhu Lake in Jiaxing City and faces a cross the lake to the administrative cultural center and old commercial and financial center. This scheme focuses on the planning concept of creating "urban lake and green chain" and follows the purpose of creating pluralistic compound urban waterfront public center and modern waterfront travel service center, to build this waterfront block into a vigorous center integrating various functions, such as leisure, entertainment, travel service, commercial and business office, creative cultural development, artistic performance, expo and exhibition, etc. The planning forms an overall layout structure of "one center, one axis and one ring" to build a landscape layout of "water belt, green corridor, blue net" and establish F-shape truck highway network system as traffic support.

杭州未来科技城重点建设区域城市色彩规划
Urban Color Planning at Key Construction Area of Hangzhou Future Sci-Tech City Project

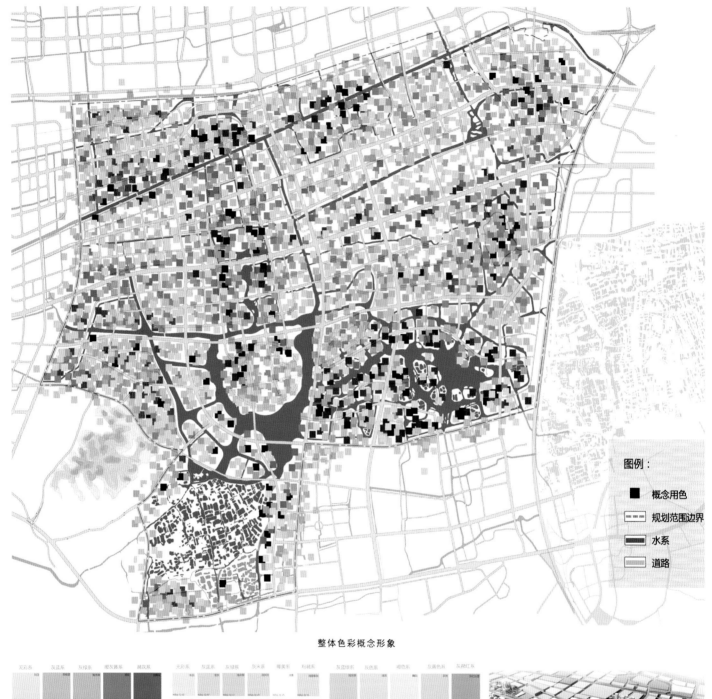

整体色彩概念形象

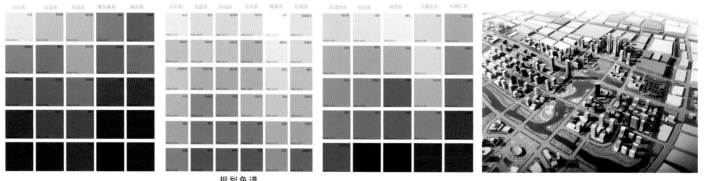

规划色谱

本案位于杭州主城区西侧，毗邻西溪湿地。本案在"如何塑造智慧城市的时代感和科技感"、"如何营建和谐生态的湿地氛围"与"如何传承地域人文特色"三条线索的指引下，采用"提取、演绎、归纳"的策略，最终形成"水韵墨章·逸彩智谷"的整体色彩概念形象，构建了"智慧、生态、人文"的城市色彩意象。

This project is located in western urban area of Hangzhou and is adjacent to the Xixi Wetland. Based on three clues including "how to create contemporary appearance and technological quality of an intelligent city", "how to maintain a harmonious ecological wetland environment" and "how to inherit regional humanism characteristics", this scheme adopts the strategy of "abstraction, deduction and conclusion" to result an integral color concept image of "waterfront landscape • colorful intelligent valley" and an "intelligent, ecological and humanism" urban color image.

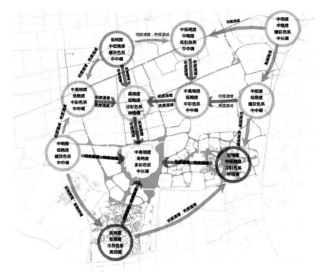

色彩"主旋律"结构

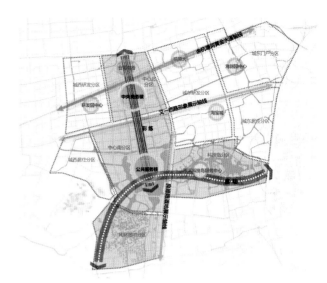

城市色彩空间结构

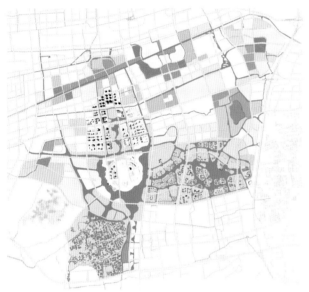

墙面色彩规划

屋面色彩规划

建设单位：杭州未来科技城管理委员会	Owner: Hangzhou Future Sci-Tech City Management Committee
功能用途：科技新城	Function: New Sci-Tech City
设计年份：2013年	Design Year: 2013
建设地点：浙江省杭州未来科技城	Construction Site: Hangzhou Future Sci-Tech City, Zhejiang
规划用地：38 km²	Planning Land: 38 km²

栖霞滨湖新城控制性详细规划及重点地段城市设计
Controlled Planning Details and Urban Design of Important Locations of Qixia Waterfront New City Project

1 游船码头
2 度假酒店
3 内湖公园
4 水主题馆
5 沿街商业
6 步行景观带
7 休闲娱乐设施
8 望湖广场
9 步行商业街
10 商业娱乐
11 商业商务办公
12 生态山体
13 天桥
14 休憩广场
15 花卉游园
16 滨湖商业
17 商务办公
18 商业金融
19 购物中心
20 酒店
21 行政办公
22 高层住宅
23 初中/小学
24 多层住宅
25 市政府大楼
26 市府广场

建设单位：山东栖霞市建设局
功能用途：居住、商业、金融、度假
设计年份：2010年
建设地点：山东省栖霞市长春湖
规划用地：946 hm²

Owner: Municipal Construction Bureau of Qixia, Shandong
Function: Residence, Commerce, Finance and Resort
Design Year: 2010
Construction Site: Changchun Lake, Qixia City, Shandong
Planning Land: 946 hm²

项目位于山东烟台栖霞市长春湖东侧。规划基于高效紧缩、多元复合、生态网络、和谐共生的理念将滨湖新城定位为以行政商务、休闲居住、金融服务、休闲度假为主导功能，以文化体育、休闲娱乐为互补功能的多元功能复合"水城核心区"。

在运用GIS地理信息技术进行用地综合评价的基础上，实施"沿湖、东控、南拓、北伸"的空间发展战略，最终形成"一带、三心、三轴、多核、四片区"的生态网络化组团布局结构，使滨湖新城成为既相对独立又与老城区相协调的新型城区。

This project is located by east bank of Changchun Lake in Qixia City. The planning focuses on the concept of effective, compact, pluralistic, compound, ecological, latticed and harmonious coexistence, to build the waterfront new city into a pluralistic and multifunctional "waterfront urban center" integrating various functions, such as administrative business, leisure residence, financial service, resort, cultural sport and entertainment, etc.

After general evaluation of the land by using GIS technique and implementation of the space development strategy of "developing along the lake, controling the eastern portion, exploring the southern portion, expanding the northern portion", an ecological network layout structure of "one belt, three centers, three axes, multiple cores and four zones" is finally formed to build the waterfront new city into a relatively independent and harmonious new urban district.

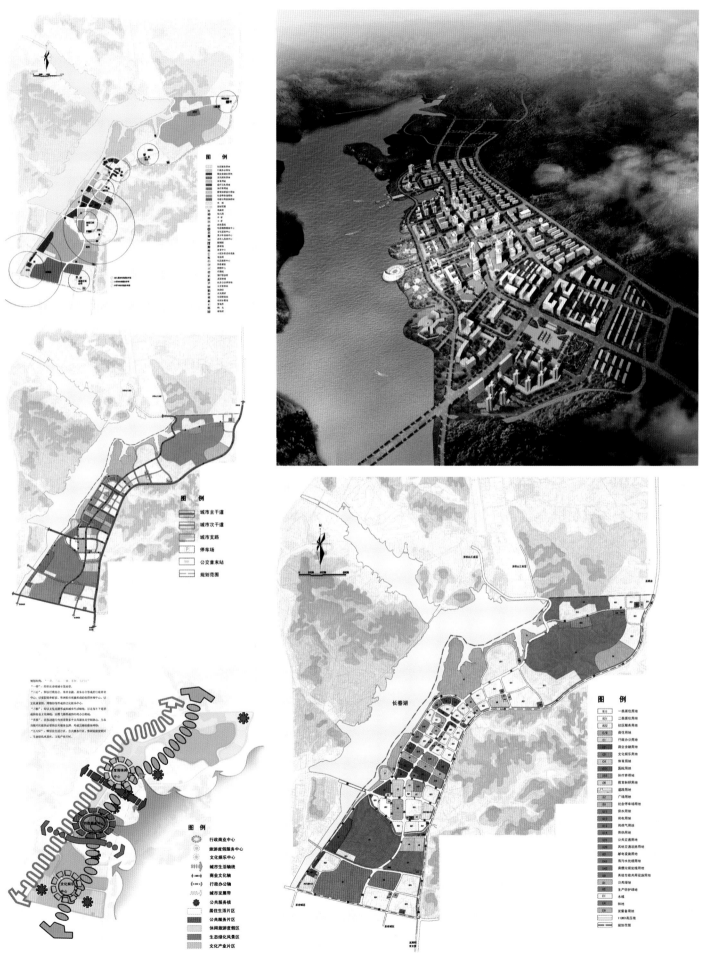

诸暨市枫桥镇枫江文化体育公园景观设计
Landscape Design of Fengqiao Town Fengjiang Cultural and Sports Park, Zhuji

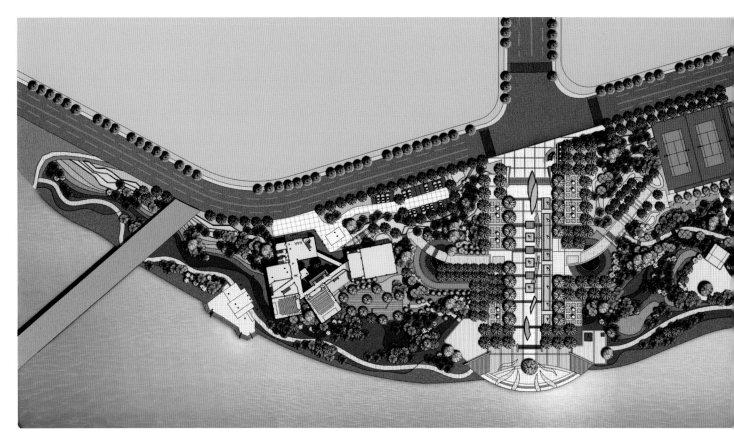

本项目是枫桥镇的主要文化体育建设项目，我们从满足运动休闲开放的城市绿地功能、发挥景观生态效益及尊重土地延续文脉三方面出发，营造一处充满阳光与活力的开放型城市公共绿地，充分发挥其城市体育文化公园的重要性，从而提升整个城市的整体环境品位。

枫桥文化体育公园景观分为A、B、C、D、E五个区块，分别为文化体育中心、公园休闲主题广场、室外运动场地、儿童游戏区域、群艺广场，组成了整个枫桥文化体育公园的景观结构框架，是集丰富的自然景观、体育活动和生态健身于一体的主题公园。

This project is a main sports construction project in Fengqiao Town. The scheme focuses on three aspects including satisfaction of the sports, leisure and open urban green land functions, realization of ecological benefit in landscape and respect of land context, to create an open urban public green land full of sunshine and vigor, to give full play to the importance of urban sports and cultural park and to increase the overall environmental quality in the city.

Landscape in Fengqiao Cultural and Sports Park is divided into A, B, C, D and E plots, which are cultural and sports center, park thematic square, outdoor sports field, children zone and collective square respectively. This forms the landscape structure of the whole park and results a theme park integrating abundant natural landscape elements, sports activities and ecological sports, etc.

建设单位：诸暨市枫桥镇人民政府
功能用途：体育公园
设计/竣工年份：2012年/～
建设地点：浙江省诸暨市枫桥镇
规划用地：54 400 m²

Owner: Fengqiao Town People's Government, Zhuji
Function: Sports Park
Design/Completion Year: 2012/～
Construction Site: Fengqiao Town, Zhuji City, Zhejiang
Planning Land: 54,400 m²

幕墙设计
Curtain Wall Design

298	绍兴金沙半岛酒店
300	杭州新天地E地块
301	诸暨市规划展示馆和科技馆
302	环球万豪国际中心
303	海亮大厦
304	海威银泰喜来登酒店
305	杭州中威电子股份有限公司安防监控设备生产基地
306	温州市江滨商务区CBD片区17-03#地块
307	杭州新天地商务中心P4地块
308	中信银行杭州分行总部大楼
309	杭州钱江新城勇进中学
310	上海长风8号东地块项目
311	梦工场影视道具生产基地
312	青山湖越秀城市综合体A区酒店
313	乐清南虹广场

298	Shaoxing Jinsha Peninsula Hotel
300	Hangzhou New World Plot E
301	Zhuji Planning and Exhibition Hall and Science and Technology Hall
302	Huanqiu Marriott International Center
303	Hailiang Tower
304	Haiwei-Intime Sheraton Hotel
305	Security Monitoring Equipment Production Base of OB Telecom Electronics Co., Ltd.
306	Wenzhou Jiangbin Business District CBD Plot 17-03
307	Hangzhou New World Business Center Plot P4
308	China CITIC Bank Hangzhou Branch Headquarters
309	Hangzhou Qianjiang Yongjin Middle School
310	Shanghai Changfeng No.8 East Plot
311	Dreamworks Film and TV Property Production Base
312	Hotel in Qingshanhu Yuexiu Urban Complex Plot A
313	Yueqing Nanhong Plaza

绍兴金沙半岛酒店
Shaoxing Jinsha Peninsula Hotel

该建筑主要采用玻璃幕墙、铝板幕墙等形式。

主楼玻璃幕墙采用隐框玻璃幕墙与竖向大明框相结合，采用了无副框幕墙系统，本系统铝合金含量低，安全可靠，具有较高的性价比。幕墙包边采用了铝单板幕墙，形成波浪形的边缘。裙楼的主入口锯齿造型由不同斜面的玻璃幕墙组成，增加了设计难度。建筑采用框架式幕墙体系，龙骨为钢型材，玻璃为中空超白LOW-E玻璃。

This building mainly adopts glass curtain wall and aluminium plate curtain wall.
The glass curtain wall of main building combines hidden frame supported glass curtain wall and longitudinal large exposed frame supported curtain wall together and uses subframe-free curtain wall system, which has low content of aluminium alloy, favorable safety and reliability, high performance-cost ratio. Single aluminium plate curtain wall is used to form waving edges. Main entrance to the podium building is designed in sawtooth shape composed of glass curtain walls with different inclined surfaces, and such kind of curtain wall also produces great difficulties to the design. Frame supported curtain wall system is made of profile steel joist and hollow ultrawhite LOW-E glass.

建设单位：绍兴华昌酒店管理有限公司	Owner: Shaoxing Huachang Hotel Management Co., Ltd.
设计/竣工年份：2014年/～	Design/Completion Year: 2014/～
建设地点：浙江省绍兴市柯南大道以北	Construction Site: North of Ke'nan Boulevard, Shaoxing City, Zhejiang
幕墙面积：21 000 m²	Curtain Wall Area: 21,000 m²
总建筑高度/层数：96 m/24 F	Total Height/Floor: 96 m/24 F
建筑设计：同济大学建筑设计研究院（集团）有限公司	Architectural Design: Tongji Architectural Design (Group) Co., Ltd.

杭州新天地E地块
Hangzhou New World Plot E

建设单位：杭州汇庭投资发展有限公司
设计/竣工年份：2012年/~
建设地点：浙江省杭州市下城区
幕墙面积：26 000m²
总建筑高度/层数：64m/17F
建筑设计：浙江大学建筑设计研究院

Owner: Hangzhou Huiting Investment Development Co., Ltd.
Design/Completion Year: 2012/~
Construction Site: Xiacheng District, Hangzhou City, Zhejiang
Curtain Wall Area: 26,000 m²
Total Height/Floor: 64 m/17 F
Architectural Design: Architectural Design and Research Institute of Zhejiang University

本项目建筑为竖向U形建筑体形。幕墙分为5个系统：明框玻璃幕墙系统、石材幕墙系统、防火玻璃幕墙系统、轻钢玻璃雨篷系统、铝板幕墙系统。

明框玻璃幕墙的横竖龙骨均采用铝合金闭口型材，在保证结构安装的前提下，大大减小铝合金龙骨截面，从而减轻重量。微晶石幕墙，采用背栓式铝合金挂件，横竖龙骨同样采用铝合金型材。防火玻璃幕墙，在保证建筑立面不被破坏前提下，确保防火分区的安装。

The building is a longitudinal U-shape volume, and it has five different curtain wall systems, including exposed frame supported glass curtain wall system, stone curtain wall system, fireproof glass curtain wall system, light steel glass awning system and aluminium plate curtain wall system.
Both transverse and longitudinal joists of exposed frame supported glass curtain wall are made of closed profiled aluminium alloy to guarantee that cross section and weight of aluminium alloy joist are obviously reduced with smooth structural installation. Micro-crystal stone curtain wall uses back bolted aluminium alloy hanger and its transverse and longitudinal joists are made of profiled aluminium alloy. Fireproof glass curtain wall could guarantee fireproof segment installation without damage of building facade.

诸暨市规划展示馆和科技馆
Zhuji Planning and Exhibition Hall and Science and Technology Hall

建设单位：诸暨城市建设投资发展有限公司
设计/竣工年份：2013年/~
建设地点：浙江省诸暨市
幕墙面积：28 800 m²
总建筑高度：31 m
建筑设计：同济大学建筑设计研究院（集团）有限公司

Owner: Zhuji Urban Construction and Investment Development Co., Ltd.
Design/Completion Year: 2013/~
Construction Site: Zhuji City, Zhejiang
Curtain Wall Area: 28,800 m²
Total Height: 31 m
Architectural Design: Tongji Architectural Design (Group) Co., Ltd.

本工程形态简洁大方且由双层表面组成，外层铝板幕墙底部与顶部采用收分处理，顶部及立面不同标高处被挖空的洞口采用GRC面板干挂饰面；内层幕墙由圆形窗不规则摆列构成。幕墙系统主要分为：双层穿孔铝板幕墙、GRC幕墙、大跨度玻璃幕墙等。

双层正六边形铝板单元板块由7根直径250 mm铝圆管焊接，且边部组装公母料附框，然后固定在折形立柱上；立面GRC幕墙做开缝设计，且地面同样做GRC面板；开孔T形钢立柱做14 m跨度的玻璃幕墙；圆形窗为中旋开启。

The building has simple but generous appearance and it has double skins. The top and bottom of outer aluminium plate curtain wall are contracted and the holes excavated at top and different levels of facade are decorated with dry-hanging GRC panel; the inner curtain wall is composed of irregularly positioned round windows. Main curtain wall systems used in this project include: double-layer perforated aluminium plate curtain wall, GRC curtain wall, wide-span glass curtain wall, etc.
Double-layer regular hexagonal aluminium plate unit is made of 7 pieces of welded 250 mm aluminium round tube, its edges are assembled with male/female attached frame to fix it on a zigzag column; slots are cut on facade GRC curtain wall and GRC panel is also used to decorate the ground; 14 m-span glass curtain wall is installed on perforated T steel column; hinged circular windows are used.

环球万豪国际中心
Huanqiu Marriott International Center

建设单位：浙江环球房地产集团有限公司
设计/竣工年份：2012年 / ~
建设地点：浙江省绍兴市迪荡新城
幕墙面积：85 000 m²
总建筑高度/层数：210.8 m / 41 F
建筑设计：浙江工业大学建筑规划设计研究院有限公司

Owner: Zhejiang Huanqiu Real Estate Group Co., Ltd.
Design/Completion Year: 2012/~
Construction Site: Didang New City, Shaoxing City, Zhejiang
Curtain Wall Area: 85,000 m²
Total Height/Floor: 210.8 m / 41 F
Architectural Design: Architectural and Planning Design and Research Institute Co., Ltd. of Zhejiang University of Technology

　　该工程塔楼为八边形，幕墙采用隐框玻璃系统，裙楼采用明框玻璃幕墙、石材幕墙、铝板幕墙。
　　1000系统为隐框幕墙，采用8（Low-E）+16A+8中空钢化玻璃；2000系统玻璃同1000系统；3000系统为避难层横向玻璃百叶系统，采用8+1.52PVB+8钢化夹层玻璃；4000系统为主楼弧形推拉窗系统，采用5（Low-E）+9A+5钢化中空玻璃；5000系统为主楼钢结构连廊幕墙系统，采用6（Low-E）+12A+6钢化中空玻璃；6000系统为裙楼明框幕墙系统，采用8（Low-E）+12A+8钢化中空玻璃；7000系统为裙楼石材幕墙系统，采用不锈钢石材干挂形式；8000系统为吊顶铝板和压顶铝板系统。

This project is an octagonal tower, whose upper floors use hidden frame supported glass curtain wall system and the podium building uses exposed frame supported glass curtain wall, stone and aluminium plate curtain wall. 1000 system is hidden frame supported curtain wall made of 8 (Low-E)+16A+8 hollow tempered glass; 2000 system is same as 1000 system; 3000 system is transverse glass shutter system for refuge storey and it is made of 8+1.52PVB+8 tempered laminated glass; 4000 system is arched sliding window system for main building and it is made of 5(Low-E)+9A+5 hollow tempered glass; 5000 system is curtain wall system for steel structure overhead corridor of main building, and it is made of 6(Low-E)+12A+6 hollow tempered glass; 6000 system is exposed frame supported curtain wall system for podium building, and it is made of 8(Low-E)+12A+8 hollow tempered glass; 7000 system is stone curtain wall system for podium building, and it is made of stainless steel dry-hanging stone; 8000 system is aluminium plate system for suspended ceiling and coping.

海亮大厦
Hailiang Tower

建设单位：杭州海亮置业有限公司
设计/竣工年份：2014年/~
建设地点：浙江省杭州市滨江区
幕墙面积：23 000 m²
总建筑高度：127 m
建筑设计：浙江工程设计有限公司

Owner: Hangzhou Hailiang Real Estate Co., Ltd.
Design/Completion Year: 2014/~
Construction Site: Binjiang District, Hangzhou City, Zhejiang
Curtain Wall Area: 23,000 m²
Total Height: 127 m
Architectural Design: Zhejiang Engineering Design Co., Ltd.

本建筑为隐框玻璃幕墙、石材幕墙、金属幕墙组合形式，采用框架式幕墙体系，幕墙最高点标高127.2 m。
主楼玻璃幕墙采用隐框铝合金骨架，玻璃为中空钢化超白LOW-E玻璃。幕墙开启窗为内开内倒，开启窗为800mm×2200mm。石材幕墙采用无焊接石材干挂系统，横梁与立柱为专用龙骨，较之常规钢通及角钢，重量更轻，之间采用螺栓连接，连接角码上设有长圆孔，保证横梁的伸缩性能。该系统可以缩短现场安装时间，同时避免现场焊接作业，消除火灾安全隐患。主楼顶部采用曲面仿石材铝板。

This tower uses a composite curtain wall system of hidden frame supported glass curtain wall, stone curtain wall, metal curtain wall and adopts frame supported curtain wall system. The highest elevation of the curtain wall is 127.2 m.
Glass curtain wall of the main building uses hidden frame supported aluminium alloy joist and hollow ultrawhite tempered LOW-E glass. The curtain wall is equipped with 800 mm×2,200 mm inward tilt-turn windows. Stone curtain wall uses weld-free stone dry-hanging system. Its beams and columns are made of special joist, which is lighter than ordinary steel tube and angle steel. They are connected by bolt, and slotted hole is set on the connecting angle to guarantee the expansion performance of beam. This curtain wall system could effectively shorten the field installation schedule, avoid filed welding and eliminate potential fire hazard. The roof of main building is made of curved stone and aluminium plate.

海威银泰喜来登酒店
Haiwei-Intime Sheraton Hotel

建设单位：杭州海威房地产开发有限公司
设计/竣工年份：2013年 / ~
建设地点：浙江省杭州市滨江区
幕墙面积：45 000 m²
总建筑高度/层数：160 m / 45 F
建筑设计：中国联合工程公司

Owner: Hangzhou Haiwei Real Estate Development Co., Ltd.
Design/Completion Year: 2013/~
Construction Site: Binjiang District, Hangzhou City, Zhejiang
Curtain Wall Area: 45,000m²
Total Height/Floor: 160 m/45 F
Architectural Design: China United Engineering Corporation

该建筑由裙楼和主楼组成，裙楼三层，塔楼四十五层，地下三层。该建筑采用玻璃幕墙、石材幕墙、单点式玻璃幕墙等形式。

主楼玻璃幕墙采用无附框玻璃幕墙系统，该系统铝合金含量低，安全可靠，具有较高的性价比。裙楼主入口的钻石造型由不同斜面的玻璃幕墙组成，加大了设计难度。建筑采用框架式幕墙体系，龙骨为钢型材，玻璃为中空超白LOW-E玻璃。

The building is composed of 3-storey podium building, 45-storey upper tower and 3-storey basement. Various curtain wall systems are used in this building, such as glass curtain wall, stone curtain wall and single-point supported glass wall, etc.
The upper tower uses frame-free glass curtain wall system, whose aluminium alloy content is low and which is safe and reliable and has high performance—cost ratio. Main entrance to the podium building is designed in diamond shape composed of glass curtain walls with different inclined surfaces, and such kind of curtain wall also produces great difficulties to the design. Frame-type curtain wall system is made of profile steel joist and hollow ultrawhite LOW-E glass.

杭州中威电子股份有限公司安防监控设备生产基地
Security Monitoring Equipment Production Base of OB Telecom Electronics Co., Ltd.

建设单位：杭州中威电子股份有限公司
设计/竣工年份：2013年/～
建设地点：浙江省杭州市滨江区
幕墙面积：19 000 m²
总建筑高度/层数：86 m/20 F
建筑设计：浙江大学建筑设计研究院

Owner: OB Telecom Electronics Co., Ltd.
Design/Completion Year: 2013/～
Construction Site: Binjiang District, Hangzhou City, Zhejiang
Curtain Wall Area: 19,000 m²
Total Height/Floor: 86 m/20 F
Architectural Design: Architectural Design and Research Institute of Zhejiang University

　　本项目建筑整体为竖向圆柱体形。幕墙分为四个系统：隐框玻璃幕墙系统、微晶石幕墙系统、轻钢玻璃雨篷系统、隐框玻璃幕墙采光顶系统。
　　弧形隐框玻璃幕墙，铝合金横梁采用后安装法闭口型材。微晶石幕墙采用背栓式铝合金挂件，横竖龙骨均采用铝合金。

The building has a longitudinal column volume, and it has four different curtain wall systems, including hidden frame supported glass curtain wall, micro-crystal stone curtain wall system, light steel glass awning system and hidden frame supported glass curtain wall skylight system.
The arc-type hidden frame supported glass curtain wall uses post-installation beam made of closed profiled aluminium alloy. Micro-crystal stone curtain wall uses back bolted aluminium alloy hanger and its transverse and longitudinal joists are made of profiled aluminium alloy.

温州市江滨商务区CBD片区17—03#地块
Wenzhou Jiangbin Business District CBD Plot 17-03

建 设 单 位：温州市国资投资集团有限公司
设计/竣工年份：2012年 / ~
建 设 地 点：浙江省温州市江滨商务区
幕 墙 面 积：28 000 m²
总建筑高度/层数：69 m /15 F
建 筑 设 计：温州市建筑设计研究院

Owner: Wenzhou State-Owned Assets Investment Group Co., Ltd.
Design/Completion Year: 2012/~
Construction Site: Jiangbin Business Zone, Wenzhou City, Zhejiang
Curtain Wall Area: 28,000m²
Total Height/Floor: 69m/15F
Architectural Design: Wenzhou Architectural Design and Research Institute

　　本建筑采用隐框玻璃幕墙、明框玻璃幕墙、石材幕墙、点式幕墙、索结构幕墙、屋顶钢构、金属幕墙组合形式，采用框架式幕墙体系。
　　主楼玻璃幕墙采用隐框铝合金骨架，玻璃为中空钢化超白LOW-E玻璃。石材幕墙采用无焊接石材干挂系统，横梁与立柱采用专用龙骨，较之常规钢通及角钢，重量更轻，之间采用螺栓连接，连接角码上设有长圆孔，保证横梁的伸缩性能。主楼南立面5楼到14楼大空间采用索结构全通透的玻璃幕墙，为现代感极强的都市风格。主楼最顶部7 m高度全部采用钢结构桁架，一圈连通后外侧采用石材幕墙外包。

The building adopts composite frame supported curtain wall system composed of hidden frame supported glass curtain wall, exposed frame supported glass curtain wall, stone curtain wall, point supported curtain wall, cable supported curtain wall, roof steel structure and metal curtain wall, etc. Glass curtain wall of the main building uses frame-hidden aluminium alloy joist and hollow ultrawhite tempered LOW-E glass. Stone curtain wall uses weld-free stone dry-hanging system. Its beams and columns are made of special joist, which is lighter than ordinary steel tube and angle steel. They are connected by bolt, and slotted hole is set on the connecting angle to guarantee the expansion performance of beam. 5-14 floors of the main building adopt large space design, and the south facade is cable supported glass curtain wall, producing vivid modern urban building features. The top 7 meters of the main building are connected by a circle of steel truss structure and are wrapped in stone curtain wall.

杭州新天地商务中心P4地块
Hangzhou New World Business Center Plot P4

建 设 单 位：杭州灵锐投资发展有限公司
设计/竣工年份：2012年/~
建 设 地 点：浙江省杭州市东新街
幕 墙 面 积：15 000 m²
总建筑高度：54.5 m
建 筑 设 计：杭州市建筑设计研究院有限公司

Owner: Hangzhou Lingrui Investment Development Co., Ltd.
Design/Completion Year: 2012/~
Construction Site: Dongxin Street, Hangzhou City, Zhejiang
Curtain Wall Area: 15,000 m²
Total Height: 54.5 m
Architectural Design: Hangzhou Architectural Design and Research Institute Co., Ltd.

　　本建筑采用玻璃幕墙与穿孔铝板幕墙组合形式，采用框架式幕墙体系，幕墙最高点标高54.45 m。主楼玻璃幕墙采用半隐框铝合金骨架，玻璃为中空钢化超白LOW-E玻璃。在幕墙玻璃外侧采用竖向400 mm×70 mm大装饰条作为立面遮阳系统。裙楼为双层幕墙，内幕墙采用隐框铝合金骨架，玻璃为中空钢化超白LOW-E玻璃。外幕墙采用穿孔铝板，里外幕墙间距有600 mm，确保立面遮阳效果，又不影响室内采光。

The building uses composite frame supported curtain wall system of glass curtain wall and perforated aluminium plate curtain wall, whose highest elevation is 54.45 m. Glass curtain wall of the main building uses semi-hidden frame supported aluminium alloy joist and hollow ultrawhite tempered LOW-E glass. Longitudinal 400 mm×70 mm large decorative strips are used as sunshade system outside the glass curtain wall. Podium building adopts double-layer curtain wall, of which the inner curtain wall is made of hidden frame supported aluminium alloy joist and hollow tempered ultrawhite LOW-E glass and the outer curtain wall is made of perforated aluminium plate. 600 mm spacing is reserved between these two layers of curtain wall to guarantee sunshade effect on facade without affecting lighting effect.

中信银行杭州分行总部大楼
China CITIC Bank Hangzhou Branch Headquarters

建设单位：杭州华凯房地产有限公司
设计/竣工年份：2011年/～
建设地点：浙江省杭州市解放东路
幕墙面积：36 000 m²
总建筑高度/层数：99.6m /20F
建筑设计：英国Foster+Partners\华东建筑设计研究院有限公司

Owner: Hangzhou Huakai Real Estate Co., Ltd.
Design/Completion Year: 2011/～
Construction Site: Jiefang East Road, Hangzhou City, Zhejiang
Curtain Wall Area: 36,000 m²
Total Height/Floor: 99.6 m/20 F
Architectural Design: Foster+Partners\East China Architectural Design and Research Institute Co., Ltd.

　　本项目建筑整体呈回字形，幕墙面积约36 300 m²，其中外幕墙为26 500 m²，内幕墙8 200 m²，屋面系统1 600 m²。内外幕墙建筑造型均为钻石形，有五种角度的幕墙面。幕墙分为分为六个系统。1000系统为室外透明单元式幕墙，标准板块为3 m×5 m；2000系统为主入口构件式系统，最大玻璃板块达到1.4 m×10 m；3000系统为室外非透明单元式幕墙，最大单元板块为3.3 m×5 m；4000系统为单元式金属板幕墙；5000系统为中庭构件式防火幕墙；6000系统为屋面系统，包含玻璃采光顶、中庭开启天窗、金属装饰条幕墙、太阳能光电采光顶、铝板屋顶、铝合金格栅、擦窗机系统等。

In this project, the building is positioned in a 回-shaped plane layout and has a total curtain wall area of about 36,300 m², including 26,500 m² external curtain wall, 8,200 m² internal curtain wall and 1,600 m² roof system. The internal and external curtain wall systems are designed in diamond shape and five different angles. There are six different curtain wall systems, among which 1000 system is unitized outdoor transparent curtain wall system, whose unit panel is 3 m×5 m; 2000 system is frame-supported glass curtain wall system for main entrance, whose maximum unit glass panel is 1.4 m×10 m; 3000 system is outdoor opaque unitized curtain wall system, whose maximum unit panel is 3.3 m×5 m; 4000 system is unitized metal sheet curtain wall system; 5000 system is frame-supported fireproof curtain wall system for atrium; 6000 system is roof system, including glass transparent roof, openable atrium skylight, decorative metal strip curtain wall, transparent solar photovoltaic roof, aluminium roof, aluminium alloy grid and window cleaning system, etc.

杭州钱江新城勇进中学
Hangzhou Qianjiang Yongjin Middle School

建设单位：杭州市钱江新城建设指挥部
设计/竣工年份：2012年/~
建设地点：浙江省杭州市钱江路
幕墙面积：33 000 m²
总建筑高度/层数：48 m /13 F
建筑设计：杭州市建筑设计研究院有限公司

Owner: Hangzhou CBD Construction Office
Design/Completion Year: 2012/~
Construction Site: Qianjiang Road, Hangzhou City, Zhejiang
Curtain Wall Area: 33,000 m²
Total Height/Floor: 48 m /13 F
Architectural Design: Hangzhou Municipal Architectural Design and Research Institute Co., Ltd.

　　本工程包含一栋行政楼单体，通过局部架空的裙楼，与具有复杂造型的教学综合楼及图书馆、风雨操场联系在一起，组成了学校南面及西面的教学功能区。
　　幕墙面积约33000 m²，其中陶板面积25000 m²，明框及隐框玻璃幕墙8000 m²。
　　玻璃幕墙采用明隐框通用系统，兼容性强，安装方便，同时采用闭口横梁，与立柱使用插销连接，方便施工。

In this project, an administrative building is linked through locally-lifted podium building to general teaching building of complex shape, library and covered playground, forming a teaching functional zone at south and west parts of the campus.
This project has a total curtain wall area of about 33,000 m², including 25,000 m² ceramic panel curtain wall and 8,000 m² exposed frame and hidden frame glass curtain wall.
General curtain wall system for exposed frame and hidden frame is used to realize high compatibility and convenient installation. Closed beam is pinned with columns, providing more conveniences for construction.

上海长风8号东地块项目
Shanghai Changfeng No.8 East Plot

建 设 单 位：上海雅戈尔置业有限公司
设计/竣工年份：2012年/～
建 设 地 点：上海市普陀区
幕 墙 面 积：120 000 m²
总建筑高度/层数：71.9 m / 22 F
建 筑 设 计：上海天华建筑设计研究院有限公司

Owner: Shanghai Youngor Real Estate Co., Ltd.
Design/Completion Year: 2012/~
Construction Site: Putuo District, Shanghai
Curtain Wall Area: 120,000 m²
Total Height/Floor: 71.9 m / 22 F
Architectural Design: Shanghai Tianhua Architecture Design Company Limited

本项目包含十栋高层住宅和一栋会所。幕墙工程主要为石材幕墙，石材选用葡萄牙砂岩。

This project consists of 10 high-rise residential buildings and one chamber. Stone curtain wall made of Portugal sandstone is used as main curtain in this project.

梦工场影视道具生产基地
Dreamworks Film and TV Property Production Base

建 设 单 位：杭州梦工场影视文化有限公司
设计/竣工年份：2013年/~
建 设 地 点：浙江省杭州市钱江经济开发区
幕 墙 面 积：41 000 m²
总建筑高度/层数：86.6 m /12 F
建 筑 设 计：华优建筑设计院

Owner: Hangzhou Dreamworks Film and TV Culture Co., Ltd.
Design/Completion Year: 2013/~
Construction Site: Qianjiang Economic Development Zone, Hangzhou City, Zhejiang
Curtain Wall Area: 41,000 m²
Total Height/Floor: 86.6 m/12 F
Architectural Design: Huayou Architectural Design Institute

    本工程包含一栋L形单体高层与加工车间裙楼，主楼为三个按"品"字分布的圆筒，由弧形线条的幕墙将其联系在一起。幕墙面积约41 000 m²，其中铝板幕墙面积10 000 m²，玻璃幕墙及室外玻璃栏杆31 000 m²。铝板幕墙采用全螺栓连接钢龙骨，为满足立面效果要求，铝板横向做凹凸线条。玻璃幕墙铝合金龙骨采用闭口横梁，与立柱使用插销连接，方便施工。

This project consists of one L-shape single high-rise building and processing workshop podium building. The main building is composed of three circular cylinder volumes, which are distributed in triangular layout and are connected by arch curtain wall. The curtain wall covers a total area of about 41,000 m², among which 10,000 m² is aluminium plate curtain wall and 31,000 m² is glass curtain wall and outdoor glass handrail. Aluminium plate curtain wall is bolted on steel joist. To realize the required effect, aluminium plate is prepared into transverse wave shape. The blocked aluminium alloy joist is used for glass curtain wall and is pinned to column, providing much convenience for construction.

青山湖越秀城市综合体A区酒店
Hotel in Qingshanhu Yuexiu Urban Complex Plot A

建设单位：杭州越秀房地产开发有限公司
设计/竣工年份：2012年 / ~
建设地点：浙江省杭州市临安经济开发区
幕墙面积：37 000 m²
总建筑高度/层数：132.2 m / 31 F
建筑设计：杭州市建筑设计研究院有限公司

Owner: Hangzhou Yuexiu Real Estate Development Co., Ltd.
Design/Completion Year: 2012/~
Construction Site: Lin'an Economic Development Zone, Hangzhou City, Zhejiang
Curtain Wall Area: 37,000 m²
Total Height/Floor: 132.2 m / 31 F
Architectural Design: Hangzhou Architectural Design and Research Institute Co., Ltd.

酒店位于省级重点工程青山湖科技城中心位置，幕墙面积约37 000 m²，整体建筑造型方正，按1.5m间距布置的铝合金装饰线条，采用浅银色作外观处理，上部竖直，下部采用弯弧造型，并布满整个建筑物，产生江南丝绸的飘逸感，整个酒店完全采用同一种隐框玻璃幕墙系统，最大玻璃尺寸4.2 m×1.5 m，设计师在原具有专利的无附框隐框系统上增加了隐蔽式安全夹具，保证了玻璃的安全性能。主楼幕墙玻璃外侧椭圆形铝合金竖向装饰型材经过弯圆处理后与立柱型材固定，实现了建筑师的设计构思。

This hotel project is located in the center of Qingshanhu Scientific and Technological City, a provincial important project of Zhejiang. Its curtain wall covers an area of about 37,000 m² and the whole building is a square volume decorated with light silver aluminium alloy strips with strict 1.5 m spacing, whose upper end is vertical and lower end is bent, like a huge piece of silk flying in the sky. The whole volume uses the same hidden frame glass curtain wall system and the largest piece of glass is 4.2 m×1.5 m. Designer has added concealed clamp to the patented hidden frame system without sub-frame, to guarantee the glass safety. The architect's design concept is realized by bending the vertical decorative oval profiled aluminium alloy elements outside the glass curtain wall of main building and fixing them to the column.

乐清南虹广场
Yueqing Nanhong Plaza

建设单位：嘉恒置业有限公司
设计/竣工年份：2012年/~
建设地点：浙江省温州市乐清
幕墙面积：260 000 m²
总建筑高度/层数：258.7 m /32 F
建筑设计：PCAL帕莱登/中国建筑上海设计研究院有限公司

Owner: Jia Heng Real Estate Co., Ltd.
Design/Completion Year: 2009/~
Construction Site: Yueqing, Wenzhou City, Zhejiang
Curtain Wall Area: 260,000 m²
Total Height/Floor: 258.7 m /32 F
Architectural Design: PCAL/ China Shanghai Architectural Design and Research Institute Co., Ltd.

项目位于浙江省乐清市，由超高层酒店、高层住宅和SOHO、多层大型商业、地下两层停车库以及三栋小型商业和配套用房组成。幕墙总面积约260000 m²，包含住宅部分的石材幕墙、铝合金门窗；商业部分的石材幕墙、玻璃幕墙、玻璃采光顶；酒店部分的玻璃幕墙、石材幕墙、铝板装饰条、雨篷等。

The project, located in Yueqing City, Zhejiang Province, is composed of super-high hotel, high-rise residence and SOHO, multi-storey large retail space, two-storey underground parking lot and three small retail and supporting facilities. Its curtain wall covers a total area of about 260,000 m², including stone curtain wall and aluminium alloy doors and windows of residential building; stone curtain wall, glass curtain wall and glass skylight of retail space; glass curtain wall, stone curtain wall, decorative aluminium strip and awning of hotel, etc.